THE
WORLD OCEAN
an introduction to oceanography

WILLIAM A. ANIKOUCHINE

Consultant Oceanographer
Santa Barbara, California

RICHARD W. STERNBERG

Associate Professor of Oceanography
Department of Oceanography
University of Washington
Seattle, Washington

PRENTICE-HALL, Inc., Englewood Cliffs, New Jersey

PRENTICE-HALL INTERNATIONAL, INC., *London*
PRENTICE-HALL OF AUSTRALIA, PTY., LTD., *Sydney*
PRENTICE-HALL OF CANADA, LTD., *Toronto*
PRENTICE-HALL OF INDIA PRIVATE LIMITED, *New Delhi*
PRENTICE-HALL OF JAPAN, INC., *Tokyo*

contents

preface   *ix*

acknowledgments   *xi*

introduction   *1*

1.1	The Principle of Unity	2
1.2	The Sea and the Environments of the Earth	3
1.3	Oceanography in the Past and Present	3
1.4	The World Ocean as a Natural Resource	4
1.5	The Future of the World Ocean	6

ONE

general features of the earth
and the world ocean   *9*

2.1	The Shape and Size of the Earth	9
2.2	The Gross Aspects of the World Ocean	10
2.3	The Depth of the Oceans	13
2.4	The Structure of the Earth	14
2.5	The Crust of the Earth	15
2.6	The Morphology of the Ocean Basins	18

TWO

origin of the continents, ocean basins,
and continental margins   *34*

3.1	Historical Review	35
3.2	Geographical Evidence	37
3.3	Sea-Floor Spreading and Plate Tectonics	43
3.4	The Origin of the Continental Margins	51

THREE

FOUR

chemical oceanography   55

4.1   Composition of Seawater   55
4.2   Sources of Seawater Substances   60
4.3   Quantity of Material Dissolved in the World Ocean   64
4.4   Dissolved Gases in the Ocean   66
4.5   Analytical Techniques of Chemical Oceanography   71

FIVE

*important physical properties
of seawater*   73

5.1   Pressure   73
5.2   Density of Seawater   74
5.3   Viscosity   79
5.4   Thermal Properties of Seawater   79
5.5   Radiant Energy and the Ocean   82

SIX

atmospheric circulation   86

6.1   Winds on an Ideal Earth   86
6.2   The Rotating Earth   88
6.3   The Effects of Continents on Atmospheric Circulation   91
6.4   Some Local Modifications   92
6.5   Diurnal Effects   93
6.6   Surface Salinity as Related to Global Wind
        Characteristics   95

SEVEN

oceanic circulation   96

7.1   A Description of the Ocean Currents   96
7.2   The Dynamics of Ocean Currents   98
7.3   The Water Masses   110

EIGHT

waves   118

8.1   Waves as Energy   118
8.2   General Features and Description of Waves   119
8.3   Classification of Waves   119
8.4   Wind Waves   121
8.5   Catastrophic Waves   127
8.6   Internal Waves   129
8.7   Standing Waves   131

tides   *133*

9.1 The General Nature of Tides   133
9.2 The Position of the Earth in the Solar System   136
9.3 The Tide-Producing Force   137
9.4 Tidal Prediction   145
9.5 Types of Tides   153
9.6 Tidal Currents   153
9.7 Measuring the Tides   158
9.8 The Tides as a Source of Power   158

NINE

inshore oceanography   *160*

10.1 The Open Coast   160
10.2 The Estuarine Environment   182

TEN

classification and description
of marine organisms   *190*

11.1 Plants in the Ocean   192
11.2 Animals in the Ocean   197

ELEVEN

introduction to biological oceanography   *212*

12.1 Development of Life in the Sea   212
12.2 Life Requirements of Marine Organisms   214
12.3 Classification of Marine Environments   219
12.4 Modes of Life in the Sea   222

TWELVE

marine ecology   *226*

13.1 Adaptations of Plant and Animal Groups to the
      Marine Environment   226
13.2 The Marine Food Web   235
13.3 The Annual Cycle of Phytoplankton   241
13.4 Some Theoretical Aspects of Food-Chain
      Relationships   244

THIRTEEN

FOURTEEN

marine sediments   246

14.1   The Components of Marine Sediments and Their
         Distribution   248
14.2   Classification of Marine Sediment   256
14.3   The Distribution of Marine Sediments   256
14.4   Variation in Marine Sediments   260
14.5   The Thickness of Marine Sediments and
         Sedimentation Rates   268

FIFTEEN

oceanographic instruments   270

15.1   Geological Measurements   270
15.2   Chemical Measurements   281
15.3   Physical Measurements   285
15.4   Biological Measurements   287

APPENDIX

physical and chemical concepts   293

A.1   Some Initial Considerations   293
A.2   Properties of Energy in the Form of Matter   297
A.3   Some Basic Chemical Definitions   309

GLOSSARY   317

INDEX   329

preface

This text is a compilation of lectures that the authors have used for several years in a beginning oceanography course for students without backgrounds in science.

A tremendous interest in using the world ocean for many purposes has been generated in recent years. This interest has accelerated the growth of the science of oceanography and has stimulated the inquiry of nonscientists. At present this country has many institutions that provide courses in basic oceanography. These courses are often crowded with individuals who have been made aware of the potential use of the sea, who are curious about the sea, or who just want to increase their general knowledge.

We have found that most students, science majors or not, are gratified if they gain an understanding of the principles of oceanographic science. It has been our experience that the principles of oceanography can be assimilated by students having practically no background in science or mathematics provided these subjects are introduced in context at the appropriate point in the course. Accordingly we have designed this text to acquaint lower-division university students with certain basic facts and the physical concepts needed to give a feeling for the general makeup of the world ocean and how it is investigated by oceanographers. Chapters are included to provide the students with a limited background in physics and chemistry as they proceed through the text. These chapters are not intended to replace the many fine lower-division physics and chemistry texts available; they reflect what the authors consider necessary so that the text can stress concepts as well as descriptions of oceanographic phenomenon. We have attempted to retain rigorousness and veracity as a necessary accompaniment to simplicity so that students proceeding in oceanography will have a valid and useful book upon which to build their knowledge.

The authors have avoided including material that usually is put in introductory texts to provide a "balanced" or "comprehensive" treatment of the subject. Historical details, repetitive examples, and non-essential facts and descriptions to which a reader is expected to "relate" or "identify" are absent. The persevering student will encounter this sort of information elsewhere at a propitious time. At the level of this text, the student would only have to labor through such passages in search of the words and phrases important enough to underline with his colored pencil. It is our intent that a judiciously rubricated copy of this text contain virtually all underlined words.

A person having little or no scientific background can study this text with the aid of a qualified instructor and gain not merely a myriad of facts and figures about the world ocean but some feeling for the sea as an entity, a "consistent thing," that is shaped and controlled by its surroundings and in turn modifies all with which it is in contact. If the student comes to realize the complexity of the world ocean and man's transitory understanding of it, one important aim of this book will have been attained.

WILLIAM A. ANIKOUCHINE
and
RICHARD W. STERNBERG

acknowledgments

Writing about the many disciplines comprising oceanography required that many people help in providing factual information, illustrations, photographs, critiques and typing services. We are grateful for this assistance and lament that we cannot list every one of the many persons involved. Still, we would like to acknowledge those who helped in a major way.

We are pleased to acknowledge the influence of Professor Richard H. Fleming of the University of Washington who is a pioneer in the field of teaching undergraduate oceanography in the United States. Although Professor Fleming had no direct contact with this book, a number of his approaches to the study of oceanography appear in its pages.

We are indebted to those institutions and individuals that permitted us to use their data and illustrations. Individual credits are given in the captions of tables and figures. Original photographic material was obtained through the courtesy of Mr. Bernard Nist, Dr. Peter B. Taylor, Dr. Billy P. Glass, Mr. Walter Brundage, Dr. William McLeish and Kent Cambridge Scientific Incorporated.

Professors John V. Byrne, Joe S. Creager, George C. Anderson, and Dora P. Henry were helpful in their reviews and discussions of the latter versions of the manuscript.

Miss Tomilynn Willits deserves a special acknowledgment for her superb typing and her perseverance, having typed all versions of the manuscript from start to finish.

A foremost acknowledgment of gratitude is extended to our wives, Joan and Lois, for their continued encouragement and patience during the preparation of this book. Their support made the job much easier.

W.A.A.

R.W.S.

THE WORLD OCEAN

ONE
TWO
THREE
FOUR
FIVE
SIX
SEVEN
EIGHT
NINE
TEN
ELEVEN
TWELVE
THIRTEEN
FOURTEEN
FIFTEEN

introduction

About 71 per cent of the earth's surface is covered by a film of water that fills a system of ocean basins that we call the *world ocean*. If the earth were the size of a grapefruit, the film of water would be the thickness of this sheet of paper. The world ocean is marked by diversity and similarity; by extremes of change that are predictable in some respects, unpredictable in many others. Many aspects are overwhelmingly complex; some are very simple. The nature of the world ocean governs much of the character of the earth and its atmosphere; and, in turn, much of the character of the ocean is determined by the earth's history, composition, and place in the solar system. The ocean was the cradle of life on this planet. In recent years, however, man himself—who evolved out of these primeval forms of life—has been modifying the ocean's character. Nevertheless, the ocean remains what poets have called it for centuries: a vast and beautiful mystery.

The study of the world ocean is called *oceanography* (from the Greek *graphos*, drawn). Oceanology (from the Greek *logos*, discourse) might be a better word but that term has not taken precedence in this country. It is difficult to separate the study of the world ocean into dis-

1

tinct descriptive, investigative, predictive, or other phases. It is just as hard to identify a purely physical aspect from a chemical aspect of oceanography. Therefore, the science of oceanography consists of many scientific disciplines brought to bear on one broad topic: the sea.

An interdisciplinary science is bound to be composed of many, often curious, fragments of the purer sciences. We find much of the description of seawater motion taken from a meteorologist's notion of winds. The study of chemistry of lakes has been adapted to learn about the fertility of seawater and its seasonal changes. Many soil bacteria live on the surface of ocean sediment; therefore, marine microbiology has roots deep in soil science. Soil mechanics, a discipline typically associated with civil engineering, has been used to study properties of materials covering the ocean bottom. The list could be extended, but it should be clear that oceanographers are persons with a variety of backgrounds who must have a substantial knowledge of most of the physical and natural sciences.

1.1   THE PRINCIPLE OF UNITY

The concept of the world ocean held by a marine scientist is somewhat different from that of the layman. All of us learn in grade school to identify the names and placement of the continents and oceans. This exercise reveals that the oceans completely surround the land masses, but it is slightly misleading because it suggests that the oceans are separated geographically. From an oceanographer's point of view, the emphasis should be on a world ocean that is completely intercommunicating. This body of water extends from the Arctic to the Antarctic; although it is forced to twist its way around the continental masses and forms distinct basins (each of which possesses a name), all of the basins are connected.

Geological oceanographers, attempting to trace the history of the earth, have discovered that the world ocean is also connected in time. The shape of the ocean basins and the position of the continental masses have changed continually for hundreds of millions, possibly billions, of years; however, their physical connection appears to span their history. Oceanographers in the 1800's and early 1900's observed the effects of these connections. In the 1870's, seawater samples collected from all ocean basins and all depths revealed striking similarities in their relative chemical content. This fact implied that the oceans were well mixed and that differences in relative proportions of various chemicals flowing in from rivers were removed by the stirring processes. Further investigation has revealed that the mixing time is about 1,000 years. It is important to remember this unity in space and time when we consider the sea as a resource, because any change that we cause in one area will eventually influence all areas.

1.2 THE SEA AND THE ENVIRONMENTS OF THE EARTH

The interaction between land, sea, and atmosphere covers such a wide range of phenomena that it is difficult to comprehend completely. Anyone who has visited a shoreline is aware of the constant struggle between land and sea. In some places, the land dominates and extends seaward in the form of river deltas, mangrove swamps, or coral reefs. In other places, the sea dominates, eroding and shaping coastal features. A shoreline reveals only one way that the sea influences man's environment. Actually, the world ocean influences almost every aspect of our lives; it affects our weather, our food and water, our recreation, international travel and commerce. The world ocean is also an important part of the global environment; because of its size and shape, it interacts with the earth's atmosphere and land masses so that it is vital to man's existence.

1.3 OCEANOGRAPHY IN THE PAST AND PRESENT

The study of the sea has changed in its emphasis, scope, and complexity throughout history. The ancient mariners traversed the sea in search of new lands or to transport goods from one port to another. Their knowledge of the sea was oriented toward winds, currents, sailing conditions, and other elements of navigation that determined the success of their voyages. In the late 19th century, however, scientists began to seek knowledge of the world ocean for its own sake.

The early scientists accomplished the first oceanwide survey of the marine environment. Probably the most famous expedition, which opened

FIGURE 1-1
The H.M.S. *Challenger* began the era of ocean exploration.

the era of ocean exploration, was made between 1872 and 1876 by *H.M.S. Challenger* (pictured in Fig. 1-1). The *Challenger*, with its crew and seven scientists, crossed the Atlantic, Pacific, and Antarctic Oceans, traveling over 125,000 km* (or 68,900 nautical miles). The expedition observed weather, currents, water chemistry at all depths, temperature, bottom topography, sediments, and marine life on a global scale. These measurements provided the factual foundation for the science of oceanography.

Nowadays, an oceanographer is often a specialist who concentrates his research effort in a specific area of study, such as biological oceanography, chemical oceanography, geological oceanography, physical oceanography, or oceanographic engineering. We have said that it is difficult to separate the study of the world ocean into phases, so naturally there is significant overlap among these categories. For example, a scientist investigating the transport of sand near the sea floor must understand both the characteristics of the sand (geological oceanography) and the nature of ocean currents (physical oceanography).

Modern oceanographic institutions house a variety of scientists whose common bond is the ocean. Their background, theoretical approach, and analytical techniques may be completely different; yet, their combined research leads toward a better understanding of the sea. These scientists use equipment that may be very sophisticated or very crude. Some of the simplest equipment includes the means for determining water clarity (a white-painted disk on a rope) and surface currents (buoyant chips or confetti sprinkled on the surface). Oceanographic research vessels are as small as rowboats or as large as ships 150 m (500 ft) long (Fig. 1-2), and there are more than 500 of them under the sponsorship of 66 nations. The rate of increase of the world's oceanographic research facilities is accelerating; yet, even with this total effort, our knowledge of the world ocean is still inadequate to solve many of the practical problems that exist (e.g., the prediction of weather, tsunamis or tidal waves, and hurricanes; the biological and geochemical effects of waste disposal).

1.4   THE WORLD OCEAN AS A NATURAL RESOURCE

The world ocean represents an incredibly large natural resource for all mankind. Although this resource is now relatively untapped, as population pressures increase, so will the demand for chemicals, minerals, food, and energy. The ocean basins contain approximately $1,370 \times 10^6$ cubic km (328 million cubic miles) of seawater, which in itself is an im-

---

* Metric units are used throughout this text. See the Appendix for a discussion of the metric system.

FIGURE 1-2
Oceanographic vessel of the University of Washington, the
Thomas G. Thompson, on an oceanographic expedition in the
Arctic Ocean. (Photograph courtesy of Joe S. Creager)

portant resource. The production of drinking water from the sea is still more expensive than naturally occurring fresh water, although desalination plants are used in water-deficient areas. Many techniques for removing the chemicals from seawater have been used, including distillation, freezing, osmosis, and ion exchange. Distillation seems to be the most economical. Approximately 700 desalination plants, each purifying over 85,000 liters (25,000 gal) of seawater per day, are in operation (or under construction) around the world, and this number will increase rapidly.

Every cubic kilometer of seawater contains about 39 million metric tons of dissolved solids, of which only common table salt (sodium chloride), magnesium, and bromine are presently being extracted in large quantities. Seawater probably contains chemical elements having a potential value of about 1 billion dollars per cubic km; however, the economic feasibility of extracting many of these elements is highly questionable because of low concentrations. Nevertheless, the world's seawater represents a large source of materials.

Other valuable ocean resources are found on or under the sea floor.

Deposits under the bottom include gas, oil, sulfur, potash, and coal. Bottom deposits include minerals precipitated from seawater, as well as sand, gravel, and oyster shells. Offshore oil and gas wells, for example, account for 17 per cent and 6 per cent, respectively, of total production in the Western world. It is expected that by 1980 one-third of the world's oil and gas will come from offshore wells. Exploitation so far has been confined to depths less than 230 m (750 ft).

Far from shore on the deep-sea floor, oceanographers have discovered and mapped vast deposits of metallic nodules that have precipitated on the bottom. Each of these nodules, called manganese nodules, is made up, on the average, of 24 per cent manganese, 14 per cent iron, 1 per cent nickel, and smaller amounts of cobalt and copper. Investigations have shown that 20 to 50 per cent of the floor of the southwest Pacific Ocean basin is covered with nodules in concentrations of 7,300 metric tons per sq km, valued at 2.35 million dollars per sq km. The total may be more than 100 billion metric tons. The chemical composition of these nodules makes it difficult to extract the manganese. However, this problem will be overcome as necessity demands.

Food resources of the sea are as great as the physical resources. The annual income of the world marine fisheries is 8 billion dollars, far in excess of the monetary return from all other types of exploitation. Of the ocean harvest, over 90 per cent is finned fish, with small percentages of shellfish and whale. Geographically, the Pacific Ocean yields 53 per cent of the marine harvest; the Atlantic Ocean yields 42 per cent; and the Indian Ocean, 5 per cent. The low figure for the Indian Ocean indicates a lack of exploitation rather than a poor source.

The marine fisheries harvest is used for several purposes, depending on the species involved. In the United States, the more desirable species are reserved for human consumption, whereas other species are converted to fish meal (as high-protein supplement for livestock and poultry) and fish oil. The development and production of fish protein concentrate (FPC) has provided a potential market for various less popular fish species and at the same time offers significant relief to protein-deficient countries. FPC is a tasteless, odorless, highly nutritious food supplement. For a cost of only 2 dollars per year, it satisfies the daily animal protein requirements of a growing child. Hopefully, this supplement will relieve serious problems of protein malnutrition that occur in some areas of the world.

1.5   THE FUTURE OF THE WORLD OCEAN

Before this century and the last one, man could exploit the sea in any way he wished without causing significant change in it. Population pressures were low and the lack of technology severely limited exploitation of the sea for food, fossil fuels, and other raw materials. Even dumping waste into the sea did not pose any serious problems, except in certain local situations.

In the 19th and 20th centuries, however, significant changes have occurred. The human population has increased alarmingly, making tremendous demands on all natural resources. Furthermore, the advance in technology has parallelled the population expansion. As a result, materials and food resources are obtainable on such massive scales that the depletion of many resources can be expected in the foreseeable future. Perhaps worst of all, the ocean is still used indiscriminately as a dumping ground for industrial and municipal wastes.

Increased quantities of lead in ocean waters has followed the use of leaded gasolines in ever-increasing numbers of gasoline engines. In some areas in fresh waters and bottom sediments, the mercury content has increased because of the increased use of fungicides and their introduction into the water. Similarly, the residues of DDT have increased too, following its development and continued use since World War II. These examples are among the more obvious observations made by oceanographers. They illustrate the point that technological advances may bring short-term advantages; but, without control, serious damage (i.e., mass destruction of biological species) could occur in only a few decades. This statement does not mean that we should discontinue our search for new products or chemicals or techniques of exploitation that will aid mankind. It suggests, however, that the evaluation of these advances should be thorough and that long-range effects are more important than short-term gains.

Experience has shown that the world ocean is not infinitely large and that it is a perishable resource. For it to continue to be useful for all peoples today and in the future, two things are necessary. First, there must be international cooperation. Serious misuse by one country will eventually affect the world ocean—and hence all countries. This situation is true whether the misuse is extreme overfishing, dumping of radioactive wastes (or any pollutant in significant quantities), or even improper oil drilling techniques that release large quantities of oil onto the sea surface. The old concept of the freedom of the seas is no longer valid. The resources of the world ocean are so valuable that they must be maintained cooperatively for the mutual benefit of all.

Second, oceanographic research must expand and continue. The findings of environmental scientists should be the key to the planned use of the sea. The food chains, paths of energy transfer, and various chemical and physical interactions are so complex that it will require the sustained effort of many scientists to monitor the sea adequately and predict future conditions. This is the challenge facing oceanographers in the immediate future.

READING LIST

DIETRICH, G., *General Oceanography*. New York: Interscience Publishers, 1963. 588p.

FAIRBRIDGE, R.W., ed., *The Encyclopedia of Oceanography*. New York: Reinhold Publishing Corp., 1966. 1021p.

HILL, M.N., ed., *The Sea*, Vols. I, II, III. New York: Interscience, 1963.

KING, C.A.M., *An Introduction to Oceanography*. New York: McGraw-Hill Book Company, 1963. 337p.

PICKARD, G.L., *Descriptive Physical Oceanography*. New York: Pergamon Press, 1968. 200p.

SVERDRUP, H.U., M.W. JOHNSON, AND R.H. FLEMING, *The Oceans, Their Physics, Chemistry and General Biology*. Englewood Cliffs, N.J.: Prentice-Hall, Inc., 1942. 1087p.

TUREKIAN, K.K., *Oceans*. Englewood Cliffs, N.J.: Prentice-Hall, Inc., 1968. 120p.

VETTER, R.C., *Oceanography Information Sources*. Washington, D.C.: National Academy of Sciences-National Research Council, 1970. 51p.

WEYL, P., *Oceanography*. New York: John Wiley & Sons, Inc., 1970. 535p.

WILLIAMS, J., *Oceanography, An Introduction to the Marine Sciences*. Boston: Little, Brown and Company, 1962. 242p.

ONE
TWO
THREE
FOUR
FIVE
SIX
SEVEN
EIGHT
NINE
TEN
ELEVEN
TWELVE
THIRTEEN
FOURTEEN
FIFTEEN

general features of the earth and the world ocean

To begin the study of oceanography, the student should understand the structure and the topographic features of the earth, especially the sea floor. Initially, oceanographers conceived the floor of the sea as a vast, undistorted plain covered with fine sediment. As sampling equipment became more sophisticated and more soundings were made, this concept changed. We now know that the floor of the sea is quite diverse in its structure and topography.

2.1   THE SHAPE AND SIZE OF THE EARTH

Our concept of the earth's shape has changed over the years. The idea that the earth is a sphere was modified when scientists realized that the earth is flattened slightly at the poles; i.e., it has nearly the shape of an oblate spheroid. This shape is the one assumed by a rotating body with equilibrium between gravitational and rotational (centrifugal) forces. The flattening causes the polar radius to be 22 km shorter than the equatorial radius.

With the advent of space exploration, scientists were able to ex-

9

TABLE 2-1

TABLE OF EARTH'S DIMENSIONS

Dimension	Magnitude	Units
Mass	6×10^{27}	kg
Volume	1.1×10^{12}	km³
Circumference	40×10^3	km
Polar radius	6,356	km
Equatorial radius	6,378	km
Total surface area	510×10^6	km²
Land surface area	149×10^6	km²
Ocean surface area	361×10^6	km²
Ocean volume	$1,370 \times 10^6$	km³
Ocean average depth	3,795	m
Ocean mean temperature	3.90	°C

amine the earth's shape from a distance. Their observations of the gravitational effects on orbiting satellites led to the conclusion that the shape of the earth is really that of an irregular, fat pear. As refined observations disclose more irregularities, it becomes obvious that the earth has a unique shape unlike any regular geometric solid. Indeed, we might say that the earth is "earth-shaped" much as we say an egg is "egg-shaped."

These small departures from the oblate spheroid shape are of some interest to oceanographers, but it is geodesists and cartographers who are most concerned with such refinements. For our purposes, it is sufficient to regard the earth as an oblate spheroid or a sphere, whichever is appropriate.

The size of the earth has been known for a long time. As early as 200 B.C., the scholar Eratosthenes calculated the earth's radius to be about 7,370 km, within 16 per cent of the accepted value. Table 2-1 gives some estimates of the earth's dimensions.

2.2   THE GROSS ASPECTS OF THE WORLD OCEAN

When a student examines the world map shown in Fig. 2-1 (inside front and back covers), he should note the following:

1. *Most of the earth's surface is covered by water.*
   The surface area of the earth is 510 million sq km. Of this area, 361 million sq km, or 70.8 per cent, is covered by water; and 149 million sq km, or 29.2 per cent, is land area. These percentages

are not constant, because there have been variations in sea level during the earth's past.* For example, during the last Ice Age in the Pleistocene epoch (2,000,000 to 11,000 years ago), enough water accumulated (as ice) on the continents to lower the sea level about 150 m. When the sea level went down, vast areas of the continental margins were exposed, and the relative proportion of land increased as much as 6 per cent. Glaciation comparable to that in the Pleistocene occurred in the late Paleozoic era and also in the late Precambrian era.

2. *Land is not distributed evenly over the globe.*
   Seventy per cent of all land is located in the northern hemisphere. There is still more water than land in both hemispheres. In the northern hemisphere, however, the ratio of water to land is 1.5 to 1; whereas, in the southern hemisphere, the ratio is 4 to 1.

3. *Continents are separated by ocean basins connected to form a single body of water.*
   The earth's six continents, Eurasia, Africa, North America, South America, Australia, and Antarctica are surrounded by the depressions of the Pacific Ocean basin, Atlantic Ocean basin (including the Arctic Ocean basin), and the Indian Ocean basin. Each ocean basin extends northward from the ocean surrounding the Antarctic continent. Because all the basins are contiguous, the system of ocean basins is called the *world ocean.*

The boundaries of the Indian, Pacific, and Atlantic Ocean basins are also shown in Fig. 2-1. By convention, the Atlantic Ocean basin is separated from the Indian Ocean basin along the 20°E meridian between the Cape of Good Hope and the Antarctic continent. Likewise, the Pacific Ocean basin is separated from the Indian Ocean basin along the 147° E meridian between Tasmania and the Antarctic continent. In contrast to these arbitrary boundaries, the boundary between the Atlantic and Pacific Ocean basins is a narrow, curved shoal, called a *sill*, that connects Cape Horn with Antarctica's Palmer peninsula. This sill delineates the Scotia Sea as a tongue of the Pacific Ocean basin extending quite far east of the southern tip of South America.

Each of the three major ocean basins has a different appearance. The Atlantic Ocean basin is the longest. It is relatively narrow and extends northward in an irregular, twisting shape. It is marked by few oceanic islands and by isolated, adjacent seas such as the Baltic Sea, North Sea, Hudson Bay, the Mediterranean Sea, and the Caribbean Sea. The

---

* A chart of geologic time is given in Table A-3.

Arctic Ocean basin is considered another sea adjacent to the Atlantic Ocean basin.

The Pacific Ocean occupies the largest ocean basin; its east-west dimension is equal to about one-half the circumference of the earth. It has a roughly symmetrical shape and contains many oceanic islands and *island arcs:* for example, the Aleutians, the Japanese islands, and the Philippine Islands. The Gulf of California and the Scotia Sea are the only adjacent seas in the eastern part of the Pacific Ocean. The western or Asiatic side of the Pacific basin, however, contains many adjacent seas, such as the South China Sea, the East China Sea, the Sea of Japan, the Sea of Okhotsk, and the Bering Sea. The world ocean is deepest in the Pacific Ocean basin. These depths are found in great trenches just seaward of the island arcs; the deepest (11,034 m) occurs in the Mariana Trench east of the Mariana island arc.

The Indian Ocean is also roughly symmetrical, but contains few islands, and is about as wide as the Atlantic Ocean basin. It lies at latitudes mostly below the Tropic of Cancer (23°N latitude). The Andaman Sea, Persian Gulf, and Red Sea, for example, are marginal to the Indian Ocean, and the Java Trench lies along its eastern edge.

There are significant differences in the area of each of the ocean basins. The Indian Ocean basin is the smallest: 75 million sq km. The Atlantic Ocean basin occupies 106 million sq km; and the Pacific Ocean basin, 180 million sq km. The area occupied by the Pacific Ocean basin is about equal to that occupied by the combined Atlantic and Indian Ocean basins. Furthermore, all of the land area of the earth's surface could fit within the borders of the Pacific Ocean basin.

In studying Fig. 2-1, we must remember that some terms used to describe geographic bodies of water have imprecise meanings. "Seas" such as the Mediterranean Sea, the Sargasso Sea, the Caspian Sea, and the Arabian Sea are geographically quite dissimilar bodies of water. The Mediterranean is virtually landlocked. The Sargasso Sea lies in the middle of the Atlantic Ocean and has only hydrographic boundaries. The Caspian Sea is a saline lake, and the Arabian Sea is merely the body of water lying on the west side of India. The term "sea" obviously has no unique connotation and should be used with caution. The term "bay" is, likewise, ambiguous. The Bay of Bengal is the body of water occupying the east side of India. The Bay of Biscay is only an irregularity in the French coast. Portage Bay is a freshwater lake in the state of Washington and is quite small in comparison with Hudson Bay, which is saline and quite large. Another of these imprecise terms is "gulf." The Gulf of California is a long arm of the sea. The Gulf of St. Lawrence is really a river mouth, or estuary. The Gulf of Mexico is a semienclosed sea.

Nevertheless, these terms are established geographic conventions that are not likely to be changed. Despite the variety of usage, they all designate areas that are but parts of a single entity, the world ocean.

TABLE 2-2

AVERAGE DEPTH OF THE OCEAN BASINS

Ocean basin	With adjacent seas	Without adjacent seas
Pacific	4,028 m	4,282 m
Atlantic	3,332	3,926
Indian	3,897	3,963
World ocean	3,795	4,117

2.3   THE DEPTH OF THE OCEANS

The average depths of the ocean basins are given in Table 2-2. It is noteworthy that the ocean basins exclusive of adjacent seas have average depths within 180 m (5 per cent) of the average for the world ocean. This fact provides the first inkling that the bottom of the ocean represents a uniform global feature. This idea is verified by the statistical distribution of depth in the world ocean and elevations on continents displayed

FIGURE 2-2

The hypsographic curve showing the percentage of the Earth's surface above a given level. The solid bars represent the percentage of the earth's surface within each particular 1000-m interval (after Sverdrup, Johnson and Fleming by permission of Prentice-Hall, Inc.).

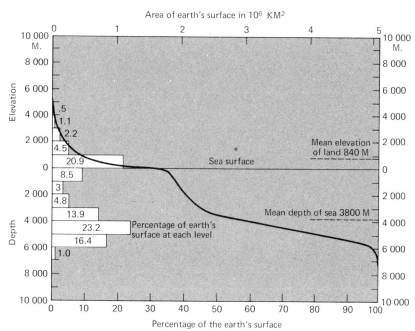

by a hypsographic curve (Fig. 2-2). A curve of this type answers the question: What percentage of the earth's surface lies above a particular level? It is not to be construed as a profile of the earth's surface. Observe from the bar graph that most of the earth's solid surface lies at two dominant levels; one lies between sea level and 1 km and represents 22 per cent of the earth's surface, and the second is the average ocean depth level lying between 3 km and 6 km. Over half of the earth's surface is at elevations or depths lying between these levels. Only about 10 per cent of the earth's surface lies at the extremes of elevation and depth.

There is a significance in the existence of the dominant levels. These are planetary features that must be considered in determining the origin and structure of the earth. Their existence is one piece of observational evidence to be included in forming hypotheses of the earth's origin and in constructing a conceptual model of its structure.

2.4   THE STRUCTURE OF THE EARTH

The model of the internal structure of the earth shown in Fig. 2-3 is based on geophysical and geochemical evidence. The earth model has three layers: (1) a *core* of iron and nickel, (2) the *mantle* of rock composed of iron silicates and magnesium silicates, and (3) an outer layer or *crust* of rock composed of aluminum silicates and magnesium silicates. The layered structure probably was formed during the formation of the earth or soon after. Some physical parameters related to these layers are given in Table 2-3.

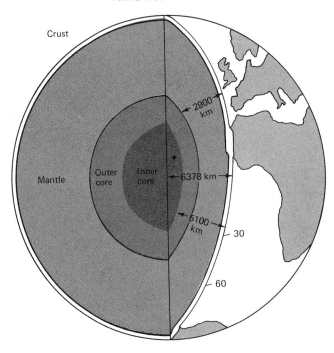

FIGURE 2-3
Structure of the earth's interior. The crustal thickness has been exaggerated so that it can be shown at this scale.

TABLE 2-3

SOME PHYSICAL PARAMETERS RELATED TO THE EARTH'S STRUCTURE

Layer	Composition	Depth Range	Approximate average density	Estimated average temperature	Relative mass
		(km)	(gm/cm³)	(°C)	(%)
Crust	Silicates of magnesium & aluminum	0–55	2.8	500	0.4
Mantle	Silicates of iron & magnesium	10–2,900	4.5	2,500	68.1
Core					
Outer	Liquid iron & nickel	2,900–5,100	11.8	5,000	⎫
Inner	Solid iron & nickel	5,100–6,370	17.0	6,000	⎬ 31.5 ⎭
Total earth			5.5		

Either of two hypotheses of the earth's origin is compatible with the three-layer earth model. According to the *tidal disruption hypothesis*, the earth originally was molten and became layered by differentiation of the earth materials as the heavier metals migrated toward the center and the lighter silicate minerals moved toward the crust and solidified.

According to the *condensation hypothesis*, however, the earth began in a relatively cool state. Differentiation of earth materials into core, mantle, and crust occurred because much heat was generated by the collision of particles and compression and by the spontaneous decay of radioactive substances. Once the temperature increased to the melting point of iron, differentiation of the core began. As the temperature increased, the silicate minerals migrated to the surface and formed the structure depicted in Fig. 2-3. The "hot origin" idea means differentiation occurred concurrently with the earth's formation; with the "cold origin" idea, differentiation covered perhaps a period of 2 billion years.

2.5   THE CRUST OF THE EARTH

Let us examine in detail the outer shell of low-density silicate rocks that form the crust of the earth. The boundary between the crust and the mantle is 10 to 55 km deep (Fig. 2-4) and is defined by an abrupt

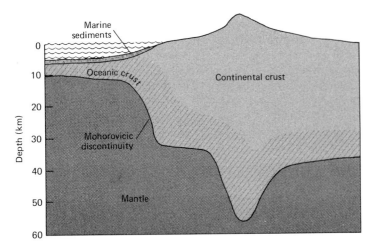

FIGURE 2-4
Idealized profile of the earth's crust.

increase in the velocity of propagation of certain seismic waves. This boundary is named the *Mohorovicic discontinuity* (or Moho) in honor of the Yugoslav seismologist who discovered the feature in 1909. Above the Moho, the earth's crust was shown by early investigations to be divided into two layers. The basis for distinguishing these layers is the velocity of propagation of seismic waves in the crust and the density of the materials in each layer. Approximately 5 km of the crust just above the Moho have properties resembling basalt (a common volcanic rock found on the earth's surface), which is composed of silicates of calcium, magnesium and iron.

The upper layer has properties similar to those of granite, a common igneous rock, composed mainly of silicates of aluminum and potassium. A more complete description of the mineral composition of these two common rock types is given in the Appendix. The arrangement of layers in the crust is shown on Fig. 2-4. The basaltic layer might extend over the entire earth but the granitic layer does not; it exists as massive continental blocks "floating" in a sea of basalt. The granitic rocks are referred to as *continental crust*; the *oceanic crust* refers to the basaltic-type rocks exposed on the sea floor. The dominant levels in the hypsographic curve (Fig. 2-2) represent the mean elevation of the tops of the continental crust blocks and the mean depth of the oceanic crust surface.

Furthermore, the dominant levels express how density differences between these two types of crust affect the configuration of the ocean basins and continents. The densities of the continental and oceanic crust are almost the same (2.8 versus 3.0); therefore, about 93 per cent of a continental block is submerged in underlying material the same way that 92 per cent of an iceberg is submerged in water. If a continent is composed of several blocks of various size, all in "floating" equilibrium, the Moho must assume a shape that reflects the surface of the continent, but with a relief nine times greater. Continental "flotation," or *isostasy* (Fig. 2-5), requires that the bottom of a continental block must rise as material is removed from the continental surface in order to keep the exposed-submerged ratio constant. This movement maintains isostatic equilibrium and is called *isostatic adjustment*.

In areas of the world that were covered by great masses of ice during parts of the Pleistocene glacial epoch, the added weight of ice caused the crustal mass to sink deeper into the mantle. When the ice melted about 11,000 years ago, this weight was removed and the crust started to rebound toward its former isostatic equilibrium position. This rebound is still happening in many places. On the Scandinavian peninsula, for example, the present rate of rebound is as much as 1 cm per year; but, as equilibrium is approached, the rate will decrease and eventually cease.

FIGURE 2-5

The principle of isostasy can be visualized by (A) flotation of ice blocks in water, or (B) flotation of land masses in "fluid" substratum.

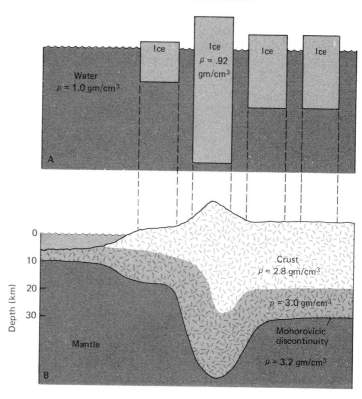

2.6   THE MORPHOLOGY OF THE OCEAN BASINS

Many of the features comprising the sea floor are very large in scale (thousands of kilometers) and occur repeatedly at certain positions within the ocean basins, thus indicating that their origin must be associated with dynamic processes active within the earth or with worldwide oceanic processes. Other features are of a smaller scale and appear to be related to local oceanic conditions. In the next chapter, we attempt to relate the configuration of the ocean basins to the earth's structure and to oceanic processes. First, however, we must describe systematically the topography of the ocean floor.

If we were to construct a series of cross-sectional profiles across the ocean basins, such as seen in Fig. 2-6, we would find some features common to all profiles. Our first observation would be the very steep nature of the slope that separates shallow from deep water at the periphery of all ocean basins. The abruptness of this slope is accentuated by the vertical exaggeration in the figure; however, it is still many times greater than most other areas of the world ocean. This slope serves as a boundary dividing

FIGURE 2-6
Outline map of the major divisions of the North Atlantic Ocean basin. Below is a representative profile from New England to the Sahara coast of Africa. The vertical exaggeration of this profile is about 40 times. (After B. C. Heezen, M. Tharp, and M. Ewing, 1959, "The Floors of the Oceans," Geol. Soc. Amer. Spec. Paper 65).

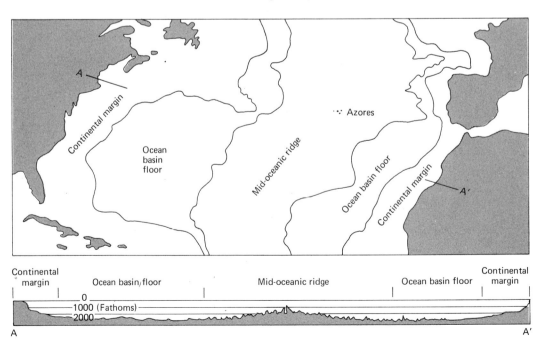

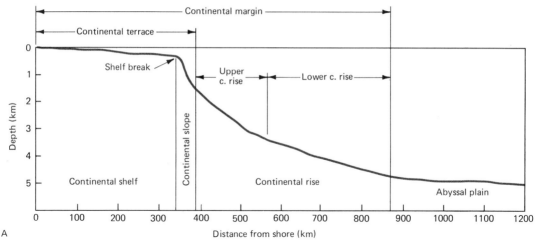

A

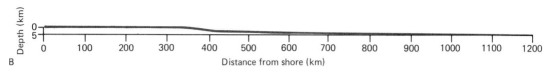

B

FIGURE 2-7
The features of the continental margin. (A) Vertical
exaggeration 150:1. (B) No vertical exaggeration.

the ocean basins into two major divisions called the *continental margin*
and the *deep sea*. These divisions serve as a focal point for our catalog of
sea-floor features.

CONTINENTAL MARGINS

The continental margin, the zone where the continents merge into
the ocean basins, is characterized by several prominent and distinct topographic features. These features are illustrated in Fig. 2-7. Surrounding
the continents of the world is a relatively flat platform called the *continental shelf*. Offshore, the inclination of this shelf increases abruptly to form
the *shelf break* or *shelf edge*. Beyond the shelf edge, the *continental slope*
extends downward to the deep-sea floor. Sometimes the shelf and slope
are lumped as a single feature and are called the *continental terrace*. In
some places, like along the margins of the Atlantic Ocean, the inclination
of the lower portion of the continental slope decreases to form what is
called the *continental rise*.

CONTINENTAL SHELVES. Continental shelves can extend as far as
1,500 km in width and 20 to 550 m in depth. On the average, the continental shelf is 78 km wide and 133 m deep at the outer edge. The average
inclination of the continental shelf is 7′ of arc or a drop of 1.8 m per km
(11 ft per mile). This inclination is too small to detect by eye.

19

TABLE 2-4

CHARACTERISTICS OF CONTINENTAL SHELVES

Nature of the Coastal Zone	Nature of Associated Continental Shelf				
	Width		Depth		
Shelves	Narrow	Wide	Deep	Shallow	
Bordering glaciated land masses	X	X			very rough topographically
Off broad coastal plains but not glaciated		X		X	very smooth platform
Associated with strong currents	X		X		coarse sediment types
Bordering large river deltas		X		X	muddy sediment
In clear tropical seas		X		X	coral reefs
Off new mountain ranges	X		X		distinct shelf may be lacking, coarse sediment

The dimensions and shape of continental shelves reflect the nature of the adjacent coast and local oceanic conditions, both past and present. Table 2-4 shows how continental shelves differ from place to place. If the coastal plain is rough and narrow, as in regions of young mountains, the continental shelf is narrow and deep or even missing. Adjacent to a broad, smooth coastal plain, such as occurs in the eastern United States, the continental shelf is broad and smooth. Where strong ocean currents impinge on the continent, the shelf is narrow, deep, and is composed of coarse sediment. At the Blake Plateau off Florida, the Gulf Stream affects the shelf in this manner. Wide, shallow shelves exist off large river deltas or behind coral and algal reefs, because sediments and biological growth are extending the continental margin seaward in these areas.

CONTINENTAL SLOPE. The continental slope is one of the major topographic features of the earth's surface. The average inclination of the continental slope is about 4 degrees of arc or about 66 m per km (400 ft per mile), but inclinations between 1 and 25 degrees of arc exist. As with continental shelves, the inclination of a continental slope can be related to effects of both geological and oceanic phenomena. The slope adjacent to

TABLE 2-5

TYPES OF CONTINENTAL SLOPES

Type	Nature of slope	Examples
Structural deformation	Steep	California
		West coast of South America
Continental rift scar	Steep	Gulf of California
Extended coastal plain and shelf	Moderate	East coast of United States
Sedimentation beyond shelf	Gentle	Gulf coast of United States
Carbonate reef (see Fig. 2-11B)	Very steep	West coast of Florida
		Yucatan, Mexico

(After R.S. Dietz, "Origin of Continental Slopes," *American Scientist*, LII, No. 1 (1964), 50–69.)

young mountain areas is generally steeper than average, ranging from 5 to as much as 25 degrees. Adjacent to large rivers that introduce great quantities of sediment, the average slope is slightly greater than 1 degree. Several types of continental slopes are described in Table 2-5.

CONTINENTAL RISE. Where observed, the continental rise is a thick prism of sediment that has been carried from the continents and deposited at the base of the continental slope. The continental rise can be as wide as 600 km, has slopes between 0.5 and 25 m per km, and lies at depths between 1,400 and 5,100 m. Sometimes the rise occurs in two distinct sections, which are labeled the upper and lower rises in Fig. 2-7.

The development of these features appears to be related to the abundance of sediment introduced from the continents. Deformation of the crust is often the result of this burden of sediment. At the present time, the Atlantic and Indian Ocean basins receive the majority of the sediment eroded from the continents, and so continental rises are developed better in these basins.

SUBMARINE CANYON. Most of the continental shelves in the world ocean are transected by *submarine canyons*: abrupt declivities that cut the continental terrace and empty onto the deep-sea floor. Submarine canyons are similar in shape and size to large canyons on land (Fig. 2-8). They are characterized by a V-shaped cross section sometimes with a flat floor, axis sloping seaward, and a sinuous course with quite accordant tributaries. Canyons cut through both unconsolidated marine sediments and hard rocks. Often they are seaward extensions of rivers that exist now or existed during the lowered sea level of the Pleistocene epoch.

The origin of submarine canyons has been debated for many years.

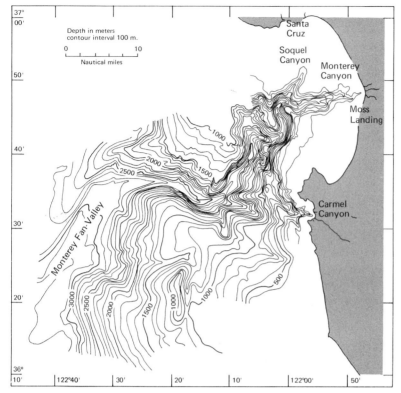

FIGURE 2-8A
Montery and Carmel canyons off central California. From soundings of
Coast and Geodetic Survey and surveys by Shephard.

FIGURE 2-8B
Comparative profiles of Monterey Canyon and the Grand Canyon using the same
number of points and same scales. The resemblance is, of course, coincidental.

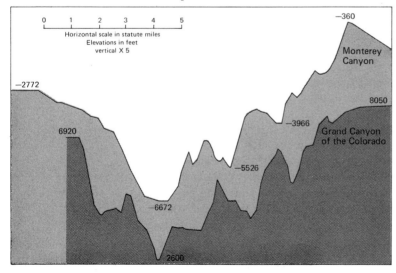

(From F. P. Shepard, Submarine Geology, 2nd Edition, Harper and Row,
New York, 1963)

It appears that rivers cut these canyons through the continental terrace when the shelf emerged during the Pleistocene. The canyons were excavated further by the action of *turbidity currents*: flows of sediment-laden water moving over the bottom of the sea. Turbidity currents have been observed directly in freshwater lakes, but their existence in the marine environment is only inferred. A series of submarine cables, cut in sequence following the Grand Banks earthquake of 1929, gave the first direct evidence that turbidity currents have erosive capabilities. Marine geologists have discovered shallow-water fauna and flora, anomalously coarse sediments, and certain sedimentary flow structures far seaward of the continental margins. These discoveries leave little doubt that turbidity currents are important mechanisms for transporting sediments from shore areas to the deep sea and that they account for widespread dispersion of sediments over a large portion of the deep-sea floor.

Vast submarine *alluvial fans* or aprons are built where submarine canyons emerge onto the sea floor. Where fans from several canyons coalesce, they form a continuous continental rise parallel to the continental periphery (Fig. 2-9). Additional material is added to the continental rise from the continental slope by submarine landslides or slumps of sediment triggered by earthquakes.

FIGURE 2-9

Block diagram of the sea floor and coast of central California. Deep-sea fans spread out from the mouths of Delgada, Monterey, and Arguello canyons. The first two fans are confined in a broad trough between the Mendocino and Murray fracture zones. (After H. W. Menard; Geological Society of America Bulletin, 1960)

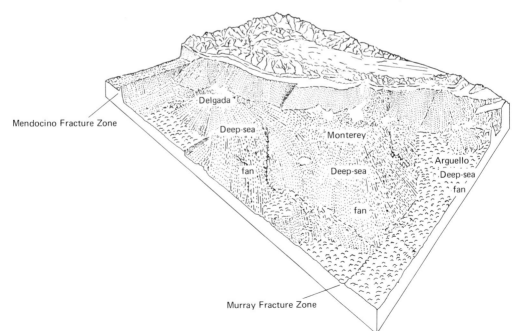

REEFS. Continental margins in tropical regions are covered with massive platforms of limestone containing the calcareous skeletal remains of plants and animals, called *coral reefs*. Since coral remains comprise only a small percentage of the total material in a reef, a better term for this feature is *carbonate reef*. Because of the life requirements for these carbonate-secreting organisms, most carbonate reefs occur in the western tropical regions of ocean basins (Fig. 2-10).

A classification of carbonate reefs proposed by Charles Darwin in 1842 and still in use defines reefs according to their proximity to shore. A reef that has been built close to shore, so that it fringes the shoreline, is called a *fringing reef* (Fig. 2-11A). The width of a fringing reef may be several hundred meters or more, but it is generally less than 30 m.

If there is a lagoon between the inner edge of the reef and the shoreline, the structure is classed as a *barrier reef* (Fig. 2-11B). Barrier reefs are generally much larger than fringing reefs. The Great Barrier Reef of Australia is a good example. It is 40 to 320 km wide and parallels the northeastern coast of Australia for a distance of 2,000 km. The lagoon depth varies from 1 m to greater than 30 m, but much of it is quite shallow. A third type of reef is the *atoll* (Fig. 2-11C) which because of its association with the deep sea is discussed in the next section.

FIGURE 2-10

Distribution of coral reefs and atolls in the world ocean. The dashed lines represent the approximate cold limit for reef coral (20°c). The diagonal lines cover the major reef areas. (Data from Putnam et al., 1960)

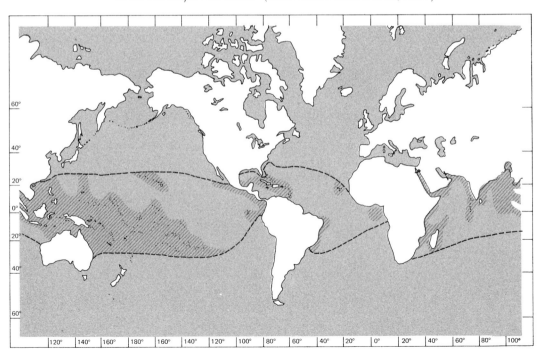

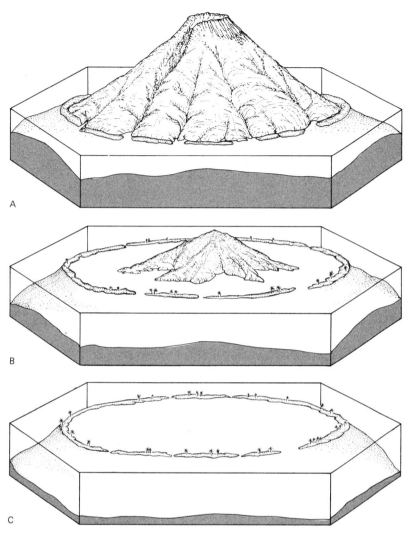

FIGURE 2-11
Types of coral reefs. (A) fringing; (B) barrier; (C) atoll. (From F. P. Shepard; *Submarine Geology*; 2nd *Edition*; Harper & Row; New York; 1963.)

DEEP-SEA FLOOR

The deep-sea floor is described in terms of the individual major features comprising it. These are abyssal plains, oceanic ridges, sea-floor fractures, deep-sea trenches, islands, and seamounts.

ABYSSAL PLAINS. Abyssal plains are broad, featureless areas of the sea floor between 3,000 and 6,000 m deep that slope less than 1 m per km (6 ft per mile). Abyssal plains usually lie adjacent to the continental margins rather than in the centers of the ocean basins. This feature is evident particularly in the Atlantic and Indian Oceans (Fig. 2-1). Approximately

25

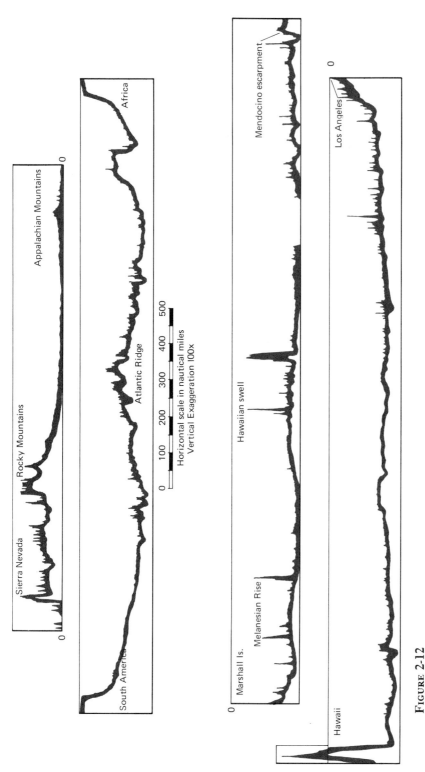

FIGURE 2-12

Comparison between a profile across the United States and profiles across the Atlantic and Pacific oceans showing the general similarity that exists between oceanic and continental topography. Vertical exaggeration times 100 for all profiles. (After F. P. Shepard; *Submarine Geology*; 2nd Edition; Harper & Row, New York, 1963).

42 per cent of the world ocean floor consists of abyssal plains. Seismic evidence indicates that abyssal plains are formed by deposition of sediment over the original sea floor. Whatever roughness existed on the sea floor when it formed has been covered or greatly smoothed. Although abyssal plains exist in the Pacific Ocean, most of the Pacific basin floor has a "hilly" topography of 200 to 400 m in relief. This hilly terrain exists in parts of all ocean basins; however, it is covered by sedimentary layers in most areas. Since relatively few large rivers drain into the Pacific Ocean, the amount of sediment introduced since its formation has not been adequate to bury these topographic irregularities.

Deep-sea drilling in the abyssal hills of the northeast Pacific has revealed that the hills are basalts that formed the sea floor prior to sedimentation there.

OCEANIC RIDGES. These ridges are mountain ranges on the sea floor that rival the size of those on the continents. The Mid-Atlantic Ridge, shown in profile in Fig. 2-12 is approximately 1,000 km wide and has rugged topography in its center and gentle foothills at its margins. Compare the profile of the Mid-Atlantic Ridge with the profile of the Rocky

FIGURE 2-13

The world-wide system of rises and ridges with related features. (After H. W. Menard from A *Symposium on Continental Drift*, 1965, by permission of the Royal Society)

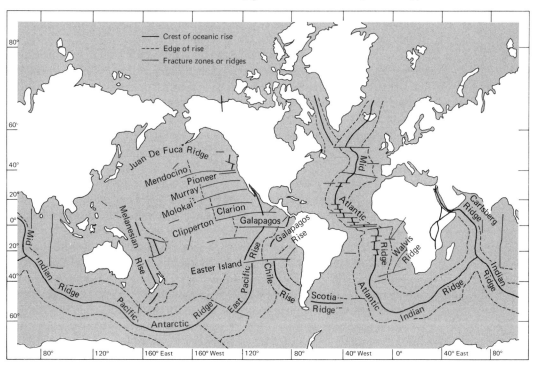

Mountains of North America, included in Fig. 2-12, and notice their gross similarities. In places, individual "peaks" rise to break the surface of the ocean and form islands. Peaks of the Mid-Atlantic Ridge form Iceland, the Azores, Ascension Island, and the island of Tristan da Cunha.

Oceanic ridges have been discovered in all ocean basins. In the north Indian Ocean and eastern Pacific basins, the ridges are extensions of surface cracks, or *fractures*, in the continental land mass. Elsewhere they form a continuous mountain chain between the continents (Fig. 2-13), so that the abyssal plains are enclosed between the oceanic ridge system and the continental rise. In almost every area where detailed bathymetric surveys have been conducted, a narrow, deep rift valley has been discovered in the central part of oceanic ridges. The rift valley is about 25 to 50 km wide and lies as much as 2,000 m below the adjacent peaks (Fig. 2-12).

SEA-FLOOR FRACTURE ZONES. The floor of the sea is cut by large fracture zones, 1 to 100 km wide, that are characterized by linear escarpments or mountainous bands, some over 3,500 km long. Fracture zones are developed prominently in the eastern Pacific Ocean basin (Fig. 2-13) where they form a series of east-west scarps with relief as great as 500 m. The scarps face both northward and southward; sometimes both kinds of scarps occur in a single fracture zone. For example, in the Atlantic Ocean basin, shorter (but still major) east-west faults offset the trend of the Mid-Atlantic Ridge along most of its length. The Carlsberg Ridge in the Indian Ocean and the East Pacific Rise also are cut and offset by many fractures. These *transform faults*, as they are called, are usual rather than exceptional features on the sea floor.

Sea-floor fracture zones abut the continents as do oceanic ridges. The Mendocino Fracture Zone displaces the continental shelf break near northern California and the Murray Fracture Zone continues landward as the transverse mountain ranges of southern California. There is little doubt that the major topographic features of the ocean and continents are related and form a single global system of ranges, rifts, ridges, and volcano chains.

DEEP-SEA TRENCHES. Trenches are observed along the margins of all ocean basins and are most extensive in the Pacific basin (Fig. 2-1). Most of them occur seaward of continental mountain ranges or volcanic island arcs. Examples are the Peru-Chile Trench adjacent to the Andes Mountains off South America and the Japan Trench seaward of the Japanese archipelago. Trenches are either rectilinear or arcuate, elongate depressions in the sea floor and range from 7 to 10 km deep. Their sides slope from approximately 4 degrees to as much as 45 degrees at the bottom and are often asymmetric in profile. A bathymetric chart of the Mariana Trench is shown in Fig. 2-14. This trench is approximately 70 km wide

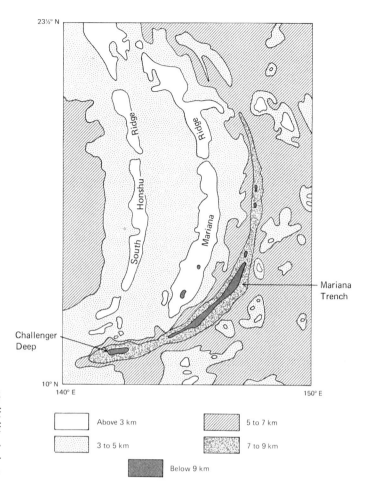

23½° N

South Honshu Ridge

Mariana Ridge

Mariana Trench

Challenger Deep

10° N
140° E   150° E

☐ Above 3 km		▨ 5 to 7 km
▫ 3 to 5 km		▦ 7 to 9 km
■ Below 9 km		

FIGURE 2-14
Detailed bathymetric chart of
the Mariana Trench area of
central western Pacific Ocean.
(Based on U.S. Navy Oceanographic office chart.)

and its length is over 2,550 km. The Challenger Deep, located in the southwest corner of the Mariana Trench, is the deepest place known in the world ocean. In 1960, the bathyscaphe *Trieste* descended 11,034 m to this point. The dimensions of some Pacific trenches are given in Table 2-6.

Trenches and troughs also occur in seas that are on the side of island arcs facing the continents. The Manila Trough, the New Britain Trench, New Hebrides Trench, Crete Trough, and Amirante Trench are examples.

Some trenches occur in parallel pairs associated with double island arcs. The Bali Trough and Java Trench off Java form such a pair. The island arc between the two depressions is nonvolcanic, whereas the inner arc is a volcanic zone. The inner trough is more shallow and less steep than the outer trench.

Another type of trench cuts oceanic ridges in oblique, transverse, or en echelon patterns. These trenches are not associated with island arcs; instead, they are related to large faults on the ocean floor. They have rec-

TABLE 2-6

DIMENSIONS OF SOME PACIFIC TRENCHES

Trench	Depth	Approximate width	Approximate length
	m	*km*	*km*
Aleutian	8,100	70	2,300
Kurile	10,542	120	2,200
Japan	9,810	100	900
Mariana	11,034	70	2,550
Philippine	10,497	60	1,400
Tonga	10,882	55	1,400
Peru–Chile	8,055	70	5,900

Data from U.S. Naval Oceanographic Office, Washington, D.C. and from Fisher and Hess, in *The Sea*, Vol. III, M.N. Hill, ed. Interscience Publishers, 1963.

tilinear trends and have symmetric profiles. The Romanche Trench in the southern mid-Atlantic is an example of this type.

SEAMOUNTS AND ISLANDS. Throughout the world ocean, irregularities rise from the sea floor. Small volcanic extrusions that rise less than 1,000 m from the sea floor are called *abyssal hills* or *knolls*. Larger volcanic extrusions that rise more than 1,000 m from the sea floor are called *seamounts*. They become *islands* if the summit broaches the sea surface. An example of a seamount that comes very close to being an island is Cobb Seamount. This volcanic cone, located approximately 325 km (180 miles) off the Washington coast of the United States (Fig. 2-15), rises from a depth of 3,000 m to within 35 m of the sea surface. This great undersea volcano is about 1.5 million years old and is part of a seamount chain extending over 1,600 km toward the Gulf of Alaska (Fig. 2-16). Frequently, seamounts occur in groups called *seamount provinces*. Some provinces, such as in the Northeast Pacific basin, contain no islands. Other seamounts are shown on Fig. 2-1.

The composition of seamounts and islands is closely related to their proximity to the continental masses. In the central part of the Pacific Ocean, the islands are composed of the basaltic-type rocks characteristic of the oceanic crust, whereas the marginal areas contain islands composed of more granitic-type rocks that are found on the continents. An imaginary line, called the Andesite Line, divides these two regions (Fig. 2-16).

Seamounts that have flat tops are called *guyots*, or *tablemounts*. They occur at all depths, and many rise to within a kilometer of the sea

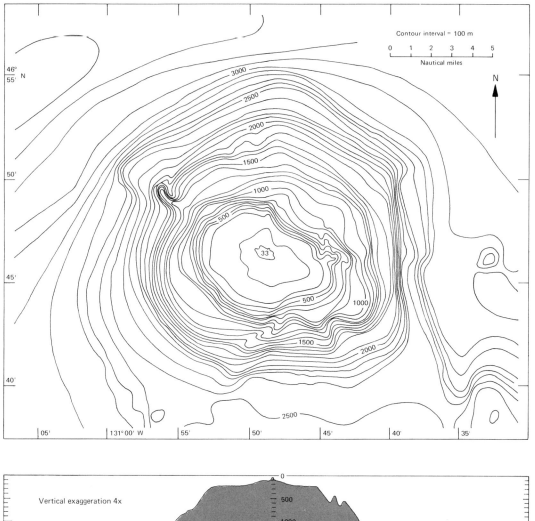

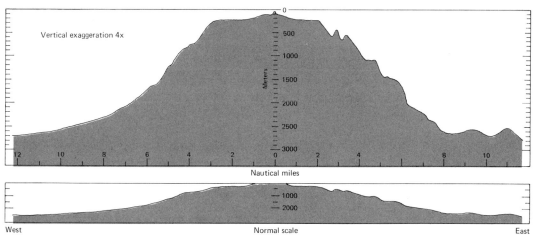

FIGURE 2-15
Plan and profile view (with and without vertical
exaggeration) of Cobb Seamount.

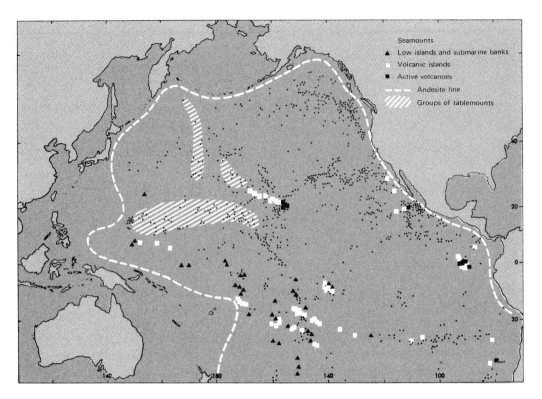

FIGURE 2-16

The distribution of observed sea mounts, tablemounts, volcanic islands, and low islands (e.g. atolls) and submarine banks in the Pacific Ocean. To date more than 2,000 such isolated features, with elevations of one kilometer or greater above the sea floor, have been discovered. There are probably at least ten times that number. (After Menard, H. W., *Marine Geology of the Pacific*, 1964, by permission of McGraw-Hill Book Company.)

surface. Tablemounts are usually found in groups or provinces; the Marshall Islands in the west equatorial Pacific are an example (Fig. 2-16). The origin of most tablemounts is volcanic; nevertheless, the flat top is difficult to explain. The top might be a wave-cut platform. But, if so, why are the tops of some tablemounts several kilometers below the sea surface and why do they appear to occur in groups? Several possibilities exist: Either the sea level rose after the waves truncated an island or group of islands, or the newly formed tablemounts sank. Possibly the tablemount never reached the surface and was always shaped that way. A flat-topped volcano in eastern Ethiopia, which appears to have formed underwater, shows no sign of wave-truncation. Instead, its shape is attributed to formation by explosions attending a submarine volcanic eruption.

In tropical areas, tablemounts and seamounts often support carbonate reefs in the form of *atolls*. An atoll is generally circular in plan and consists of a central lagoon up to 40 km in diameter and up to 75 m deep and is surrounded by a narrow carbonate reef dotted with islands

(Fig. 2-11C). Apparently an atoll forms as an island sinks, so that the rate of upward growth of the reef matches the sinking of the island or the rise of the sea level. Charles Darwin proposed this idea in 1842. According to his theory, a slowly sinking island first develops a fringing reef, then a barrier reef, and finally becomes an atoll when the original surface of the island is submerged. Drilling on atolls in the west Pacific has revealed that the structure of an atoll conforms to his theory.

READING LIST

DARWIN, C.R., *The Structure and Distribution of Coral Reefs*. Berkeley, Calif.: University of California Press, 1962. 214p.

DIETZ, R.S., "The Pacific Floor," *Scientific American*, CLXXXVI, No. 4 (April 1952), 19–23.

FISHER, R.L., AND R. REVELLE, "The Trenches of the Pacific," *Scientific American*, CXCIII, No. 5 (November 1955), 36–41.

GUILCHER, ANDRÉ, *Coastal and Submarine Morphology*. New York: John Wiley & Sons, Inc., 1958. 274p.

HEEZEN, B.C., *The Floors of the Oceans*, Vol. I, *The North Atlantic*, Geological Society of America Special Paper No. 65. New York, 1959. 122p.

KING, C.A.M., *Ocean Geography for Oceanographers*. New York: St. Martin's Press, Inc., 1963. 336p.

KUENEN, P., *Marine Geology*. New York: John Wiley & Sons, Inc., 1950. 568p.

MENARD, H.W., "The East Pacific Rise," *Scientific American*, CCV, No. 6 (December 1961), 52–61.

———, "Fractures in the Pacific Floor," *Scientific American*, CXCIII, No. 1 (July 1961), 36–41.

———, "Sea Floor Spreading, Topography and the Second Layer," *Trans. Amer. Geophys. Union*, XLVIII, No. 1 (March 1967), 217.

MORGAN, W.J., "Trenches, Great Faults, and Crustal Blocks," *Jour. Geophys. Res.*, LXXIII, No. 6 (March 15, 1968), 1959–82.

"The Ocean," *Scientific American*, CCXXI, No. 3 (September 1969), 288p.

ROBERTSON, E.C., *The Interior of the Earth, An Elementary Description*, U.S. Geological Survey Circular 532. Washington, D.C.: Government Printing Office, 1966, 10p.

SHEPARD, F.P., *The Earth beneath the Sea*. Baltimore: The John Hopkins Press, 1959. 275p.

STRAHLER, A.N., *The Earth Sciences* (2nd ed.). New York: Harper & Row, Publishers, 1971. 824p.

ONE
TWO
THREE
FOUR
FIVE
SIX
SEVEN
EIGHT
NINE
TEN
ELEVEN
TWELVE
THIRTEEN
FOURTEEN
FIFTEEN

origin of the continents, ocean basins, and continental margins

So far, we have considered only a description of the structure and surface topography of the earth. Now let us consider: (1) what caused the continents and ocean basins to assume their present shape and location; (2) when the shape and location originated; and (3) whether their shape and location has been the same throughout geologic time or has evolved slowly.

In the past few years, attempts to explain the history and origin of crustal features have been able to relate many diverse facts that heretofore could not be fitted into a coherent pattern. In fact, present ideas on *global tectonics* (the mechanics of deformation of the earth) have drawn on widespread evidence from glacial geology, biological evolutionary patterns, the shape and topography of the sea floor, and petrology of the oceanic crust. Measurements of crustal heat flow, gravity, seismic activity, and geomagnetism have also been used. Yet, many questions remain unanswered. Our understanding of the earth is better than ever before, however, and is continually changing as new facts are gathered.

3.1 HISTORICAL REVIEW

In 1915, Alfred Wegener, a German meteorologist, proposed the hypothesis that the position of the continents and the shapes of the ocean basins have not been unchanged since the origin of the earth but that the continents have been in continual motion during the past 150 million years. At that time, his idea seemed rather bizarre.

The first line of evidence for his hypothesis was the shapes of the continents. Alfred Wegener observed, as do most elementary students of geography, that the east coasts of North and South America fit the west Coast of Africa and Europe like pieces of a jigsaw puzzle. If the continents are attached, moreover, it is found that mountain ranges older than the Cretaceous period appear to fit together, whereas younger strata and mountain ranges do not bear such resemblance. Possibly, therefore, some sort of physical separation occurred about 150 million years ago.

When paleontological evidence is studied, fossils indicate that the evolutionary trends of many animals began to diverge during the lower Cretaceous period. Prior to this divergence, the evolutionary development of many animals was the same regardless of where the animals lived. In the past 150 million years, the same animal on different continents evolved differently and produced markedly different sequences of fossils. This evidence thus suggests that before 150 million years ago, continental regions of the world were somehow connected, allowing free migration of animals throughout the world. After that time, a physical separation isolated different continental regions and caused independent evolution within these isolated areas.

These observations led Alfred Wegener to propose *the theory of drifting continents.* According to this hypothesis, there was originally one large continent, which he called Pangaea, that resulted from the formation of the earth's crust (Fig. 3-1). About 150 million years ago, this continent began to break into individual blocks and drift apart in a nonuniform fashion. North and South America drifted at a rate that widened the Atlantic Ocean basin between the Americas and the Europe-Africa continental masses. Australia moved south and east from its initial position in the Indian Ocean basin.

Although continental drift did account for the evidence as presented by Wegener, many scientists disputed his interpretations. Furthermore, the continental drift theory failed to present a mechanism adequate to move continents, and so it remained in dispute until the late 1950's. However, with the advent of sophisticated geophysical measuring techniques, data on earth magnetism, seismicity, heat flow, and sea-floor topography, substantial evidence was added that tended to support this theory, or some similar one. Deep-sea drilling, begun in the late 1960's, has provided data from samples of marine sediments and underlying basement rock that has

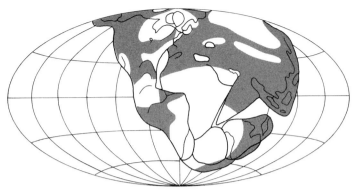

Late carboniferous

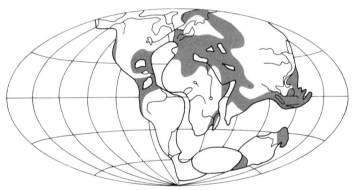

Middle tertiary

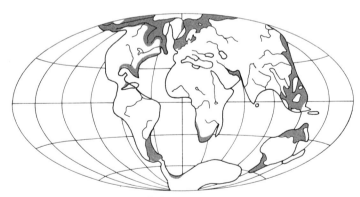

Early quaternary

FIGURE 3-1

Reconstructions of the map of the world for three periods according to Wegener's theory of continental drift. Dotted areas represent shallow seas. (From *The Origin of Continents and Oceans* by Alfred Wegener, Dover Publications, Inc., New York, 1966. Reprinted through permission of the publisher).

virtually confirmed the idea of moving continents, although the original idea of Wegener has been highly modified. The mechanism for causing these effects, however, has still to be discovered.

3.2   GEOPHYSICAL EVIDENCE

GEOMAGNETISM

Investigations of the geomagnetic record in volcanic and certain sedimentary rocks provided probably the greatest impetus in rejuvenating the continental drift theory in the 1950's. Scientists found that when a volcanic magma containing a ferromagnetic substance cools below a certain temperature in the presence of the earth's magnetic field, it becomes magnetized in such a way that its polarity has the same orientation as the geomagnetic field. Furthermore, the intensity and orientation of the rock magnetism is permanent, so it indicates the magnetic field that was prevailing when the lava solidified.

Paleomagnetic studies have shown reversals in the magnetic polarity of rocks in segments of the volcanic record, thus suggesting that the north and south magnetic poles have reversed numerous times in the past. The exact reasons for these reversals are unknown. Nevertheless, the reversal pattern has become an important tool, because it represents a series of global events that permit worldwide correlation of the age of rocks.

Figure 3-2A illustrates the magnetic orientation of volcanic formations in both hemispheres. Two worldwide reversals in the magnetic field are indicated, one ending approximately 3½ million years ago. At present, approximately 170 reversals of the earth's magnetic field are known. The reversals extend far beyond the date of 3½ million years shown in Fig. 3-2A; estimates suggest that they date back as far as 76 million years (Fig. 3-2B).

It is also possible to measure the orientation of magnetic mineral particles in marine sediments. As an iron oxide mineral (magnetite) particle settles through the water column and rests on the bottom, it tends to orient in alignment with the earth's magnetic field. When these particles are incorporated into layers of marine sediments, they preserve their magnetic alignment. Sedimentary layers on the sea floor contain a record of the same magnetic epochs that are recorded in volcanic rocks. An example is shown in Fig. 3-3.

Geomagnetic investigations at sea reveal that the magnetic intensity of oceanic crustal material near oceanic ridges follows a definite pattern. Lines of equal magnetic intensity lie parallel to the axis of the oceanic ridge system, and strips of normal and reverse magnetism are arranged symmetrically on either side of the ridge axis (Fig. 3-4). The sequence of magnetic polarity reversals on one side of the ridge is duplicated in mirror image on the other side. Hence, we observe that the crustal rocks grow

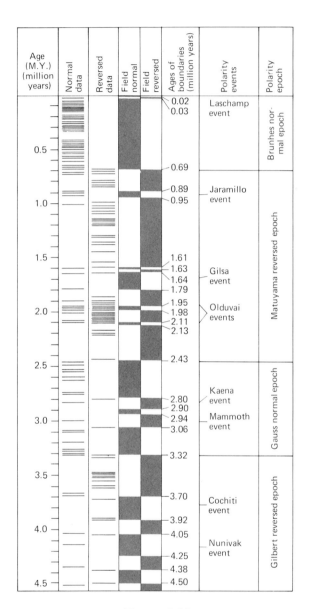

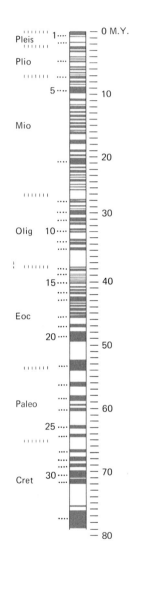

FIGURE 3-2A

Time scale for geomagnetic reversals. Normal and reversed polarity intervals are placed in chronological order based on radiometric dating of samples. (After Cox, from "Science," vol. 163, 1969, by permission of the American Association for the Advancement of Science).

FIGURE 3-2B
The geomagnetic time scale. From left to right: Time scale for geologic eras, numbers assigned to magnetic anomalies, geomagnetic field polarity (normal polarity black), age in millions of years. (From Heirtzler, el al., by permission of the American Geophysical Union).

FIGURE 3-3
The geomagnetic reversal time scale is also observed in the deep-sea sediment record. These cores were collected from the Antarctic. The gray portions represent normal polarity, the white portions represent reversed polarity. (After Heirtzler, et al., by permission of the American Geophysical Union.)

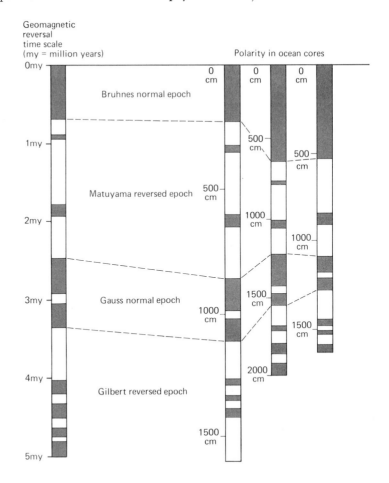

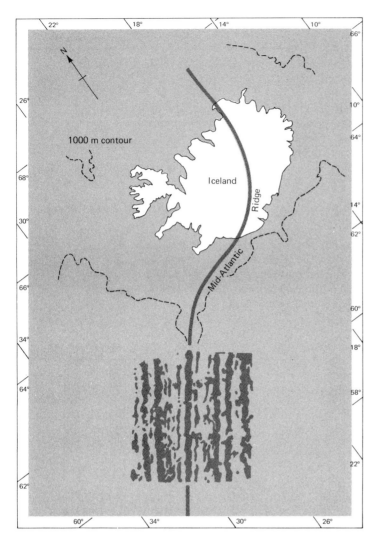

FIGURE 3-4

Anomaly pattern bordering the Mid-Atlantic Ridge in the area south of Iceland is strikingly symmetrical. The parallel bands in which the earth's field is stronger (stippled) or weaker (plain) than the regional average are oriented along the ridge's axis. The magnetic bands are presumably produced by bands of rock with normal and reversed magnetism. (After Heirtzler, et al.)

older away from the ridge and that these rocks have a particular magnetic character. The pattern is similar moving *either* direction from the ridge.

SEISMOLOGY

The first seismograph for measuring earthquake vibrations was installed in 1889. Since then, and especially in more recent decades, consider-

40

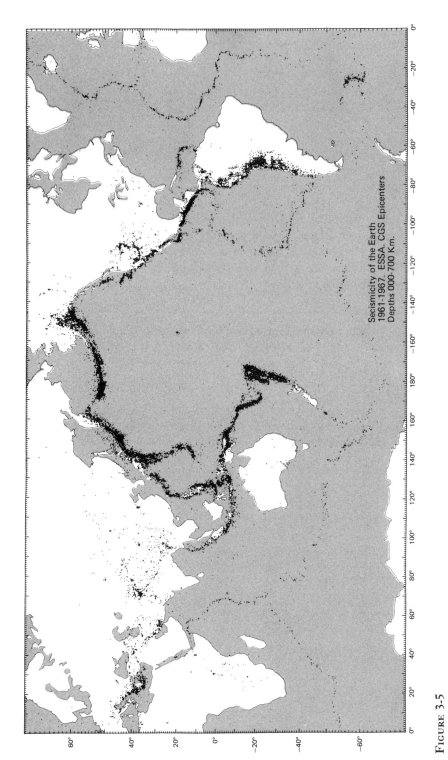

Seismicity of the Earth
1961-1967. ESSA. CGS Epicenters
Depths 000-700 Km.

FIGURE 3-5

Worldwide distribution of earthquake epicenters, oceanic trenches, and ridges. Note the association of earthquake activity with trenches and ridges.

able data have been collected on the epicenters and relative magnitudes of earth tremors occurring throughout the world.

When the distribution of epicenters is plotted on a world chart (Fig. 3-5), a rather interesting pattern emerges. Significant earthquake activity appears to be closely associated with oceanic ridges and trenches. This pattern is most evident in the deep trenches of the circum-Pacific region and the Mid-Atlantic Ridge, but an obvious correlation occurs throughout the world ocean.

Around the Pacific margin, earthquake activity occurs down to 750 km below the surface. It is well documented that the depth of earthquake foci increases in a plane dipping toward the continents from the Pacific oceanic trenches (as shown in Fig. 3-6). These data suggest some degree of differential earth movements along this plane, hence the occurrence of earthquakes.

HEAT FLOW

Precise measurements of the temperature gradient occurring in the upper few meters of the sea floor and knowledge of the thermal conductivity of the sediments allow an oceanographer to calculate the rate at which

FIGURE 3-6

Vertical section oriented perpendicular to the Tonga arc. Circles represent earthquakes projected from within 0–150 km north of the section; triangles correspond to events projected from within 0–150 km south of the section. All shocks occurred in 1965. (After Isacks et al., by permission of the American Geophysical Union.)

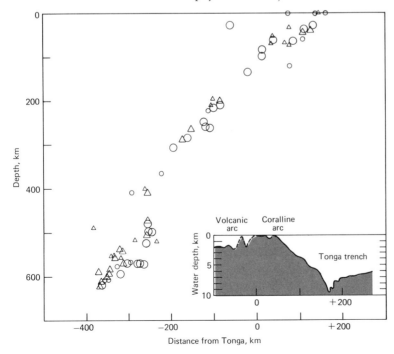

heat is being transferred through the interface. Measurements made over the past two decades have shown that a worldwide average for terrestrial heat flow is about 1.5×10^{-6} cal per sq cm per sec or 30 to 40 cal per sq cm per year. This magnitude of heat loss is about equivalent to the heat generated within the crust by the decay of radioactive elements.

Variations in terrestrial heat flow appear to follow a definite pattern. Heat flow is higher than average on oceanic ridges and island arcs and lower than average in the vicinity of oceanic trenches (Fig. 3-7).

3.3   SEA-FLOOR SPREADING AND PLATE TECTONICS

A widely held hypothesis states that the oceanic ridge system is a region where molten material from the mantle wells to the surface, cools, and spreads laterally along the sea floor. As the material cools, it becomes magnetized. This explanation would account for the linear magnetic pattern that parallels the ridge system and for the symmetrical arrangements of magnetic anomalies on both sides of a mid-ocean ridge. Called the *sea-floor spreading hypothesis,* this idea is an important breakthrough in earth science, for it correlates sea-floor morphology with the earth's internal processes and also places these features in their correct historical perspective.

In attempting to provide a mechanism for sea-floor spreading, some scientists postulate the existence of large convection cells in the upper mantle. These cells extend 750 km into the mantle, the maximum depth of earthquake foci. Above this depth, heat transfer by conduction and radiation is not sufficient to remove heat (possibly generated by radioactive decay) from the earth. The temperature gradient maintained between the upper mantle and the earth's surface is large enough to produce thermal convection of the upper mantle. The rising current of mantle material reaches the earth's surface, spreads laterally, cools, and finally sinks (Fig. 3-8). This process is analogous to a pan of water in which the heat energy added to the bottom of the pan is carried rapidly to the surface by the convective motion of the water. In the earth, however, the vertical motion is quite slow—on the order of 0.5 to 1 cm per year.

All geophysical evidence from the ocean floor tends to support the sea-floor spreading hypothesis. The ocean-ridge system drawn schematically in Fig. 3-9A shows the characteristics that would result from a rising convection current: large heat flow, earthquakes, and fracturing of the crust. Figure 3-9B depicts the region where the convected material flows under the continental block. Here the crust is bowed down forming a trench; heat flow from the crust is quite small; and earthquakes are frequent.

If the sea-floor spreading hypothesis is correct, the sea floor acts as a

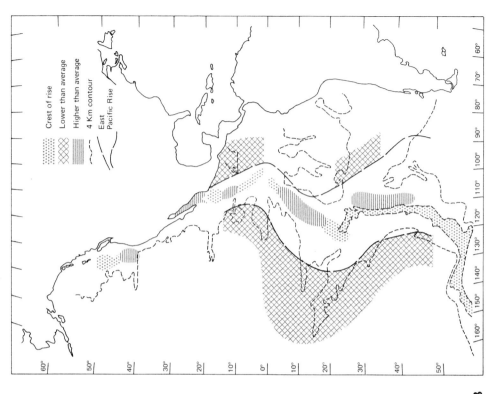

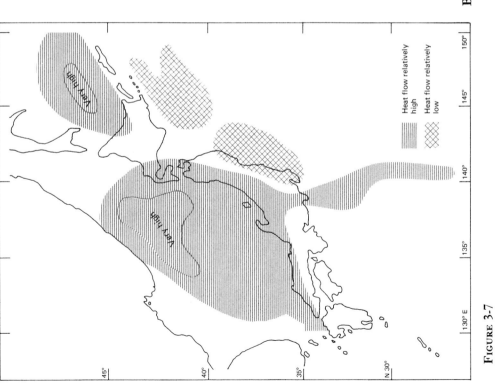

B

FIGURE 3-7

Distribution of heat flow in and around Japan (A) and the East Pacific
Rise (B). Crosshatched areas indicate regions of lower than average heat
flow and horizontally hatched area are those of relatively high heat flow.
(After Takeuchi, et al., and H. W. Menard)

A

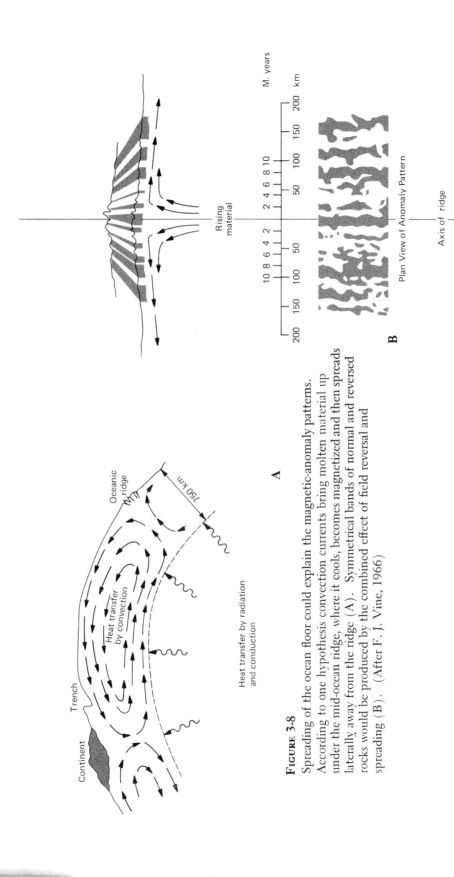

FIGURE 3-8

Spreading of the ocean floor could explain the magnetic-anomaly patterns. According to one hypothesis convection currents bring molten material up under the mid-ocean ridge, where it cools, becomes magnetized and then spreads laterally away from the ridge (A). Symmetrical bands of normal and reversed rocks would be produced by the combined effect of field reversal and spreading (B). (After F. J. Vine, 1966)

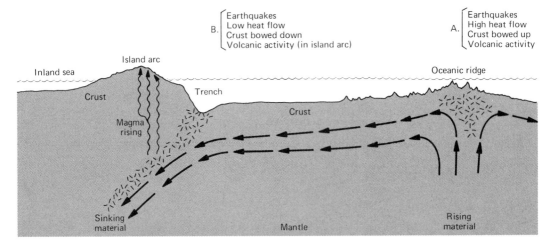

FIGURE 3-9

Idealized section through the upper mantle and crust showing the hypothesized convective motion and general characteristics of the crust in the regions of rising material (A) and sinking material (B). Regions of seismic activity are depicted by the symbol (x).

FIGURE 3-10

Schematic drawing showing several seamounts in the Northeast Pacific Basin. According to the seafloor spreading hypothesis the seamounts form on the ridge and move with the crust in the direction of spreading.

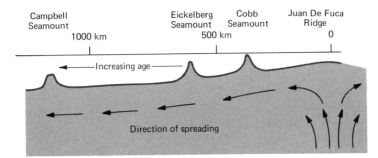

large conveyor belt moving new oceanic crust from the ocean ridges toward the trenches and carrying seamounts, oceanic islands, tablemounts, and even continents with it (Fig. 3-10). Submarine volcanoes originate in the vicinity of ocean ridges, the site of formation of new oceanic crust. An example of this movement is shown in Fig. 3-11. Islands close to the Mid-Atlantic Ridge are relatively young. For example, the Azores are 15 million years old, and Iceland is 13 million years old. At increased distance from the ridge, the islands are increasingly older. Bermuda is 35 million years old; the Faroe Islands, 25 million years old; and the Cape Verde Islands, 150 million years old. Calculations based on the age and horizontal displacement of these islands indicate that the rate of crustal movement is 1 to 8 cm per year.

Sea-floor spreading can explain why volcanic-island chains such as the Hawaiian Islands grow older in a northwestern direction, the postulated direction of crustal movement. This mechanism is shown in Fig. 3-12. If the ocean floor spreads about 8 cm per year in each direction away from the East Pacific Rise, sea-floor spreading could produce the Pacific Ocean basin in approximately 100 million years.

These speculations were verified dramatically when the cores recovered by the *Glomar Challenger* during the Deep Sea Drilling Project were analyzed. Fossils in the cored sediments showed that the initial

FIGURE 3-11

Age of Atlantic islands, as indicated by the age of the oldest rocks found in them. The numbers associated with the islands and seamounts (●) and of cores (·) gives ages in millions of years. (After J. Tuzo Wilson, 1965, by permission of the Royal Society).

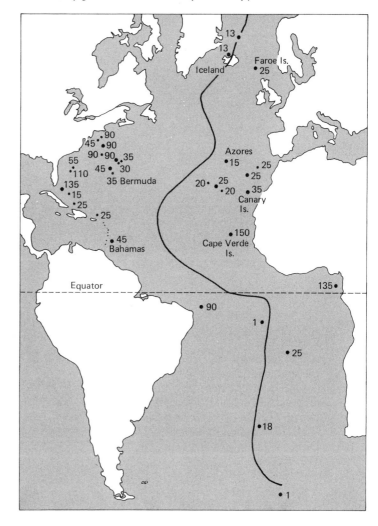

FIGURE 3-12

The arrows show linear chains of islands or seamounts which get older in the direction of the arrows (From J. Tuzo Wilson, 1965, by permission of the Royal Society)

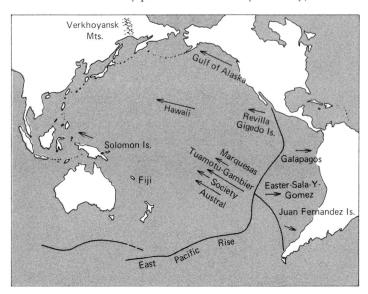

FIGURE 3-13

Age of the oldest sediments from the southern Atlantic and their distance from the ridge axis. The diagonal line indicates a spreading rate of 2 cm/yr. (After Maxwell, et al., 1970, Initial Reports of the Deep Sea Drilling Project, Vol. III, Washington.)

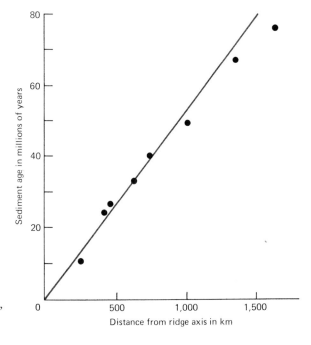

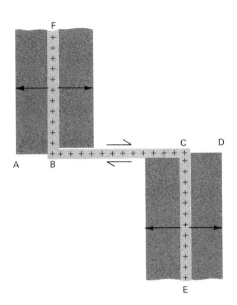

FIGURE 3-14

An idealized model of sea-floor spreading and transform faulting. Dark shading represents new surface area created along active ridge crests BF and CE. Present seismicity (crosses) is confined to ridge crests and to segment BC of the fracture zone AD. Half arrows denote sense of shear motion along BC. (After Isacks, et al., by permission of the American Geophysical Union.)

sediments (those lying directly over crustal rocks) near the Mid-Atlantic Ridge and East Pacific Rise are younger than the initial sediments at a site farther away (Fig. 3-13). The lavas that these sediments rest upon are likewise youngest near the ridge or rise and are older farther away. Furthermore, the total thickness of sediments deposited on the bottom increases with distance from the Mid-Atlantic Ridge and from the East Pacific Rise.

Spreading-rate calculations based on the deep-sea cores indicate that the South Atlantic is spreading uniformly about 1 to 4 cm per year. The Pacific used to spread 8 to 13 cm per year westward about 15 million years ago, but since then there has been an additional motion toward the north at about half that rate.

The regularity of magnetic anomalies (Fig. 3-4) suggests that the oceanic crust remains rigid as it moves. This suggestion has led to a modification of the continental drift theory called *plate tectonics*. According to this idea, the oceanic crust moves on individual plates of mantle, 70 to 100 km thick, that form at an ocean ridge and sink back into the mantle at a deep-sea trench. The plates of mantle carry rigid blocks of crust that float and conform to isostasy. The continental crust floats on these plates of mantle also; in some cases, a single plate may carry both kinds of crust on it.

Because the ocean-ridge pattern is offset in many places, fracture zones must exist where the plates move in opposite directions or move in the same direction but at different rates (Fig. 3-14). The fractures are called *transform faults*, and they are characterized by severe seismic activity.

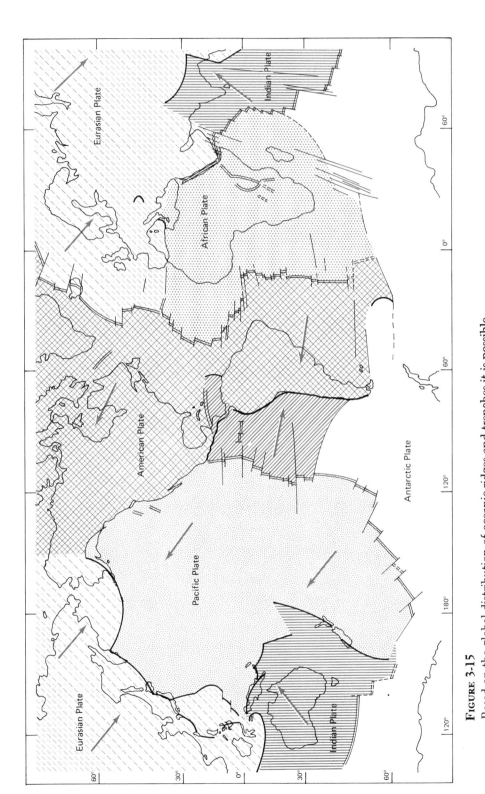

FIGURE 3-15

Based on the global distribution of oceanic ridges and trenches it is possible to divide the crust into relatively few major plates. Arrows show relative plate motion. (After Isacks, et al., by permission of the American Geophysical Union).

It is possible to recognize at least six major plates that could account for the pattern of continental drift shown in Fig. 3-15. These plates are bounded by either ridges or trenches and are about 70 to 100 km thick. Trenches occur where one plate sinks beneath another. Some continental borders lack trenches and seismic activity, because those continental masses move with the plates. For example, the plate moving westward from the Mid-Atlantic Ridge (Fig. 3-15) includes the South American continent. It rides over the eastern pacific plate along the Peru-Chile Trench, so there is no trench along the eastern border of South America.

Oceanic crust can sink at a continental border for tens of millions of years. However, if the sinking oceanic crust carries a continental block, the blocks eventually will collide. Then the process of sinking must stop or shift direction, because it is impossible for a continental block to sink in an oceanic crust of higher density. Probably, therefore, throughout geologic time, continents and ocean basins have split, rejoined, and drifted over the earth's surface. This theory may also explain the formation and position of the world's mountain ranges. There is no reason to believe that the super-continent of Pangaea lay dormant for over 4 billion years and then began to crack apart and drift 150 million years ago. These processes might have begun with the formation of the earth, with continents and ocean basins forming and reforming a number of times.

The sea-floor spreading and plate tectonics hypotheses seem to be the best ones devised so far for answering many questions regarding the history and development of the ocean basins. They relate more diverse geological information than any other hypotheses and, certainly, are the most in accord with the results of deep-sea drilling into the floor of the world ocean.

3.4 THE ORIGIN OF THE CONTINENTAL MARGINS

Early in this century, scientists believed that continental shelves were uniform in character and were simply abraded platforms cut into the continents by large waves. A shelf was said to consist of an abrasion platform nearshore and a terrace offshore on which erosional debris was deposited (Fig. 3-16). The maximum depth of the continental shelf corresponded to the maximum depth of ocean-wave activity called *wave base.*

More recent observations of the shape, structure, and sediment cover of continental shelves have greatly modified early ideas. The variety of shelf morphologies listed in Table 2-4 implies that the surface features of continental shelves are formed by several processes. A theory of the formation of the continental margins must explain the existence of rock

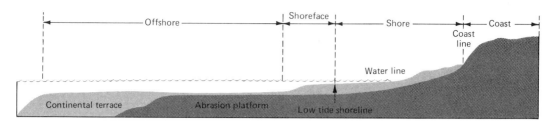

FIGURE 3-16

Profile of a graded continental shelf (After D. W. Johnson, *Shore Processes and Shoreline Development*, permission by Columbia University).

outcrops or coarse sediments at the shelf break, the presence of channels incised into the shelf, and the fact that the outer shelf often lies deeper than the influence of waves. The sea-floor spreading and plate tectonics hypotheses furnish an explanation by depicting some of the margins of the continents as geologically active.

Certainly the lowering sea level during the Pleistocene epoch was an effect common to all processes. During this time, the sea level went down to a maximum of about 130 m. As a result, many of the present shelves were exposed to subaerial erosion and deposition, which cut river-like channels into the shelf and may have supplied the coarse deposits often found at the shelf break.

Since the sea level rose at the end of the Pleistocene epoch, shelves have been modified by modern coral growth and sedimentation and have been eroded by strong currents. In regions where wave and current activity is low, the shelves are "relic" structures that reflect the Ice Age of 11,000 years ago rather than conditions today.

Geophysical profiles of the continental margins show that many of them are formed by a massive barrier lying as much as 800 km offshore of the continents. The barrier has caused the ponding of sediment eroded and transported from the continents. When these sediments fill the basin, they are transported over the barrier and form a continental slope. The extent to which the basins behind these barriers is filled with sediment depends on the sediment supply and age of the feature; all extremes have been found.

Scientists recognize several types of barriers. (1) *tectonic barriers* are the result of geological uplift or upwelling of lava (Fig. 3-17A). Tectonic activity is related to sea-floor spreading; when the leading edge of a continental plate is deformed, volcanism and faulting result. The continental slope, therefore, can begin from a fault scarp, a rift, or the surface of a mass of volcanic and possibly sedimentary rock. The example shown in Fig. 3-17A characterizes the entire Pacific rim. The tectonic barrier is seen as the Channel Islands off California, the Aleutian Islands, and the islands of Japan. (2) *Reef barriers* are caused by the growth of carbonate-secreting organisms on a preexisting shelf (Fig. 3-17B). The location of reef barriers in the world ocean is restricted to tropical latitudes where en-

52

vironmental conditions are favorable (Fig. 3-10). (3) *Diapir barriers* (Fig. 3-17C) are like those formed in the Gulf of Mexico where domes of salt pushed upward from salt deposits that are buried several kilometers deep and are 140 million years old. In regions of the world where no barriers exist, the shelf appears as in Fig. 3-17D. The eastern coast of North and South America, and much of Africa and Australia are examples.

The nature of the continental margin at any point is governed by the processes of barrier formation and sedimentation. Where sedimentation is rapid, as in deltaic areas, the margin is a depositional feature. If sedimentation is slow, the barrier structure shapes the continental margin, particularly the continental slope. In the absence of a barrier, the margin of the continent is a primary terrace mantled with a thickness of sediment. More detailed discussion of the sedimentary processes in the nearshore zone and the central and outer continental shelf are presented in Chapters 10 and 14.

FIGURE 3-17

Examples of three kinds of barriers in typical shelf cross sections; a simple shelf is also depicted. (After K. O. Emery.)

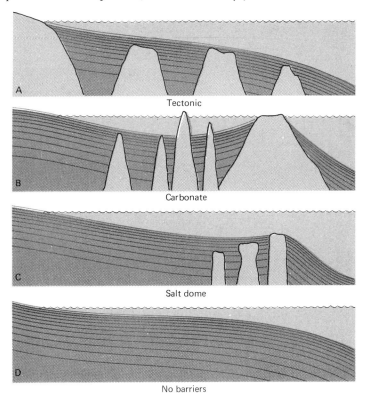

Tectonic

Carbonate

Salt dome

No barriers

READING LIST

EMERY, K.O., "The Continental Shelves," *Scientific American*, CCXXI, No. 3 (September 1969), 106–22.

ISACKS, B., J. OLIVER, AND L.R. SYKES, "Seismology and the New Global Tectonics," *Jour. Geophys. Res.*, LXXIII, No. 18 (September 15, 1968), 5855–99.

KEEN, M.J., *An Introduction to Marine Geology*. New York: Pergamon Press, 1968. 218p.

LE PICHON, X., "Sea Floor Spreading and Continental Drift," *Jour. Geophys. Res.*, LXXIII, No. 12 (June 15, 1968), 3661–97.

"The Ocean," *Scientific American*, CCXXI, No. 3 (September 1969), 288p.

PHINNEY, R.A., ed., *The History of the Earth's Crust*. Princeton, N.J.: Princeton University Press, 1968. 244p.

TAKEUCHI, H., *et al.*, *Debate about the Earth*. San Francisco: Freeman and Cooper and Co., 1967. 253p.

WEGENER, A., *The Origin of Continents and Oceans*. New York: Dover Publications, 1966, 246p.

WILSON, T.J., "Continental Drift," *Scientific American*, CCVIII, No. 4 (April 1963), pp. 86–100.

ONE
TWO
THREE
FOUR
FIVE
SIX
SEVEN
EIGHT
NINE
TEN
ELEVEN
TWELVE
THIRTEEN
FOURTEEN
FIFTEEN

chemical

oceanography

4.1   COMPOSITION OF SEAWATER

Probably every element that occurs on earth is dissolved in seawater, the dilute solution filling the basins of the world ocean. Although the study of seawater is far from complete, we know that it is composed primarily of a dozen or so elements. The remaining elements are present in only small quantities. Water is certainly the most abundant component; it makes up about 96.5 per cent of the weight of seawater.

CHEMICAL PROPERTIES OF WATER

Much of the nature of our planet is determined by the properties of water. Elsewhere in this text the influence of water's unique physical properties is demonstrated. In this section the solvent properties of water are of primary interest. The solubility of substances, especially ionic compounds, those compounds yielding ions upon dissociation in solution (see Appendix), is usually much higher in water than in other solvents. Water is able to dissolve more substances than most known liquids. The remarkable solvent properties of water arise from the atomic structure as-

55

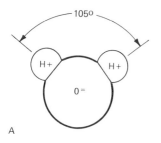

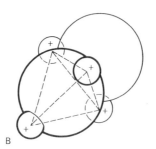

A

B

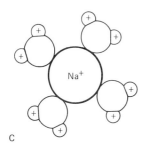

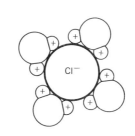

C

FIGURE 4-1

Various configurations of the water molecule. (A) The dipole arrangement of the hydrogen atoms in association with the oxygen atom; (B), The interaction of two water molecules showing the tetrahedral array of hydrogen atoms; (C), The hydration of sodium and chloride ions.

sumed by liquid H_2O. Each oxygen atom is associated with two hydrogen atoms in an array as shown in Fig. 4-1A. The two tightly bound hydrogen atoms lie on one side of the center of the H_2O group and produce a net concentration of positive charge on that side. The side away from the two hydrogen atoms has a net negative charge. The uneven distribution of charge (polarization) causes each H_2O group to act as an electric *dipole*.

Because of this dipole configuration water molecules tend to interact strongly with each other as well as other ions. The interaction between water molecules is such that a tetrahedral array of hydrogen atoms surrounds each oxygen atom (Fig. 4-1B). Two of the hydrogen atoms are bound more loosely to the central oxygen atom but are bonded strongly to an adjacent oxygen atom. This electrical bonding of water molecules causes water to have an especially high latent heat of vaporization—the heat required to separate molecules when converting liquid to gas.

Ionic compounds dissolve readily in water because water dipoles are attracted to and surround each charged ion with a sheath of oriented water molecules (Fig. 4-1C). This effect, called *hydration*, prevents recombination of dissolved ions and accounts for the large solubility of ionic salts. Non-ionic compounds are soluble in water not because they dissociate to form charged ions that can become hydrated, but because the charge distribution in these molecules is irregular and a sheath of water dipoles is attracted to sites of high charge.

MAJOR CONSTITUENTS

Table 4-1 lists the substances that comprise 100 per cent of the major constituents which in turn comprise 99.9 per cent of the elements dissolved in seawater. These major constituents exist largely as hydrated free ions. But they can exist in other forms. Small amounts of them form *ion pairs* because of electrostatic attraction between highly charged ions even in the presence of water dipoles. The sulfate ion, for example, forms ion pairs with magnesium, calcium, strontium, and other divalent ions; consequently, only 50 per cent of it exists as free ions. Some compounds dissociate only slightly. In seawater, undissociated boric acid molecules predominate (90 per cent) over borate ions (10 per cent). On the other hand, fluoride and bromide can form *complexes* with metal ions; it is likely that these complexes also exist in seawater. Strontium, however, is chemically much like calcium, so it probably exists largely as hydrated free ions.

TABLE 4-1

THE MAJOR CONSTITUENTS OF SEAWATER

Ion	Symbol	% as free ion	% by weight of the total major constituents
Cations			
Sodium	Na^+	99	30.62
Magnesium	Mg^{2+}	87	3.68
Calcium	Ca^{2+}	91	1.18
Potassium	K^+	99	1.10
Strontium	Sr^{2+}	90	0.02
Anions			
Chloride	Cl^-	100	55.07
Sulfate	SO_4^{2-}	50	7.72
Bicarbonate	HCO_3^-	67	0.40
Bromide	Br^-	100*	0.19
Borate	$H_2BO_3^-$	10*	0.01
Fluoride	F^-	100*	0.01
			100.00

* Estimated

TRACE ELEMENTS

The remaining elements in seawater are present in concentrations of less than 1 part per million. Table 4-2 lists the trace elements in the

TABLE 4-2

CONCENTRATION OF TRACE ELEMENTS IN SEAWATER EXCLUSIVE OF NUTRIENTS AND DISSOLVED GASES

Element	Symbol	Concentration	Element	Symbol	Concentration
Lithium	Li	170 ppb*	Cerium	Ce	0.005 ppb*
Rubidium	Rb	120	Yttrium	Y	0.3
Iodine	I	60	Silver	Ag	0.04
Barium	Ba	30	Lanthanum	La	0.01
Indium	In	20	Cadmium	Cd	0.1
Zinc	Zn	10	Tungsten	W	0.1
Iron	Fe	10	Germanium	Ge	0.06
Aluminum	Al	10	Chromium	Cr	0.05
Molybdenum	Mo	10	Thorium	Th	0.05
Selenium	Se	0.4	Scandium	Sc	0.04
Tin	Sn	0.8	Lead	Pb	0.03
Copper	Cu	3	Mercury	Hg	0.03
Arsenic	As	3	Gallium	Ga	0.03
Uranium	U	3	Bismuth	Bi	0.02
Nickel	Ni	2	Niobium	Nb	0.01
Vandium	V	2	Thallium	Tl	<0.01
Manganese	Mn	2	Gold	Au	0.004
Titanium	Ti	1	Protoactinium	Pa	2×10^{-6}
Antimony	Sb	0.5	Radium	Ra	1×10^{-7}
Cobalt	Co	0.1	Rare Earths		0.003–0.0005
Cesium	Cs	0.5			

* parts per billion
(From Goldberg, E.D., *Chemical Oceanography*, Vol. I. New York: Academic Press Inc., 1963.) p. 164–165.

TABLE 4-3

AVERAGE CONCENTRATIONS OF NUTRIENT IONS DISSOLVED IN SEAWATER

Dissolved element	Concentration
Nitrogen	500 ppb*
Phosphorus	70
Silicon	3,000
Carbon	Extremely variable 1,000–10,000

* parts per billion

order of their abundance in seawater. Some trace elements form free ions and ion pairs, but most of these elements exist as organic complexes and, to a lesser extent, as hydroxide, chloride, and possibly fluoride complexes.

The importance of trace elements in seawater became apparent after early attempts to prepare artificial seawater for marine aquaria failed. Even though the major and nutrient ion concentrations could be made correct, the trace element impurities in the main ingredients caused an upset in the trace element ion concentrations. This water often would not support life, so it seemed reasonable to assume that some trace element ions are important to the biology of the sea.

NUTRIENT ELEMENTS AND ORGANIC COMPOUNDS

Nutrient elements exist as nitrate (NO_3^-), phosphate (PO_4^{---}), and silicate (SiO_3^-) ions. As the name implies, the nutrient ions are the fertilizers of the sea. Although they are present in small quantities, the nutrient ions are important because they are necessary for plant growth. Approximate concentration levels and elemental ratios are given in Tables 4-3 and 4-4. Their importance will be discussed in Chapter 12, which introduces biological oceanography.

Also found in small quantities in seawater are various organic compounds that occur both as dissolved molecules and as *colloids*. These organic substances are carbohydrates, proteins and their decomposition products, lipids (fatty substances), vitamins, auxins (plant hormones), and humic substances formed by complex reactions involving the decomposition products of organisms. Humic substances are complicated compounds that have been isolated from seawater; they are also called "yellow substances." Fatty acids are particularly important because, upon incorporation in bottom sediments, they rapidly form *kerogen*, a relatively unre-

TABLE 4-4

NUTRIENT ELEMENT RATIOS (BY ATOMS)

Element	Oxygen	Carbon	Nitrogen	Phosphorus
Theoretical ratios in living organisms	276	106	16	1
Ratios in recently deposited organic sediments		96	16	
Ratios in older sediments		192	16	
Ratios observed in organic materials dissolved in seawater		32	16	

active compound that slowly yields the compounds comprising crude petroleum.

4.2   SOURCES OF SEAWATER SUBSTANCES

Let us now consider how water in the ocean was composed, how its composition is regulated, and what changes might have occurred to seawater composition throughout the history of the earth. Broadly speaking, the initial sources of the substances in the ocean must have been the primordial atmosphere, the original rock of the earth, and gases released from the earth's interior (Fig. 4-2). A small amount may have come to the ocean as cosmic dust from extraterrestrial sources.

In all probability, water has evolved from vapor released from volcanoes, fumaroles, and hot springs. Water, in combination with chemicals, has also been released during the erosion of primordial rocks. The present rate of discharge of water from the interior of the earth is about 66×10^{15} g per year. This rate can easily account for all the water on earth—provided that contributions have remained the same throughout geologic time and provided that allowance is made for considerable recycling of water through the ocean, atmosphere, rivers, lakes, and

FIGURE 4-2

Sources of sea water substances and their cycles of transportation throughout continents, atmosphere, and ocean.

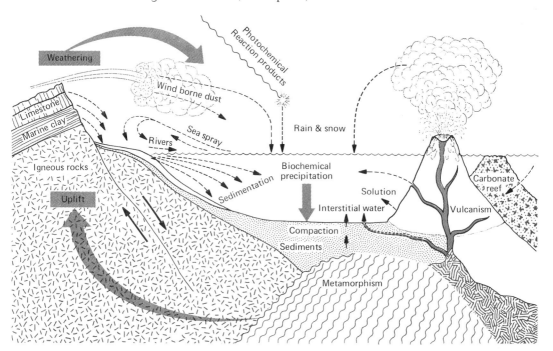

groundwater. If such assumptions are true, the total volume of water in the world ocean has been increasing steadily at about 0.4×10^{15} g per year throughout its history.

Most of the major cations (positively charged ions) and minor elements were derived from igneous rocks (see Chapter 14). This fact has been deduced by comparing the average compositions of igneous rocks, seawater, and sediments. However, to account for the chloride, sulfate, and bromide anions (negatively charged ions) in seawater, another source is required. These substances are present in volcanic discharges, so it is likely that the major anions in seawater have been added from past emissions of volcanoes.

Today, the total amount of dissolved material observed in the ocean is about 5×10^{22} g. Rivers carry about 2.5×10^{15} g to the sea annually. If we assume that dissolved substances began accumulating in the ocean about 3 billion years ago and that the rate of transport by rivers rose steadily to the present value, somehow a large excess of dissolved materials has been removed from the sea. Obviously, therefore, the sea does not trap all the ions brought to it by rivers and from other sources. Instead, there must be regulatory processes that maintain the concentrations of dissolved materials at the levels observed today. Other lines of evidence (noted in the following section) indicate that the composition of seawater has not changed much during the last 1.5 billion years. We can thus infer that the regulatory processes have been active at least that long.

PROCESSES REGULATING THE COMPOSITION
OF SEAWATER

There are several processes that regulate the chemical composition of the world ocean. When we compare the solubilities of all the elements dissolved in the ocean to the concentrations that exist, we find that just a few are present in saturation amounts. Of the major cations in the ocean, only calcium (as the bicarbonate) is near saturation; saturation is maintained by the dissolution or precipitation of solid calcium carbonate. Nitrogen and the inert, or noble, gases are near saturation; their concentrations are strongly affected by temperature.

One regulatory process is *cation exchange*. The concentrations of sodium, potassium, magnesium, rubidium, cesium, and, to some extent, calcium are controlled by cation exchange on clay minerals brought to the ocean by rivers. In river water and in the ocean, the order of preference of *adsorption* on clays is magnesium, then potassium, then sodium. Clay minerals, especially illite, incorporate potassium into their crystal structure, so this cation is less concentrated in seawater than magnesium.

The minor elements of manganese, nickel, cobalt, zinc, and copper are greatly undersaturated, because they are adsorbed on ferromanganese minerals in nodules on the ocean floor. In addition, apatite (the calcium phosphate mineral in fish-bone debris) adsorbs thorium, barium, strontium, and the rare earth elements, because, like the ferromanganese minerals, it presents active surfaces for adsorption. By these processes, trace elements are removed from seawater and concentrated in sediments and in nodules on the sea floor.

A second regulatory process is *anion exchange,* although little is known about anion exchange reactions in the ocean except that clay minerals have a substantial capacity for anions. Phosphate is taken up in sediments; there is also a relative enrichment of bromine in sediments. These conditions, therefore, may be produced by such exchange reactions.

The net behavior of all processes controlling the concentration of an element dissolved in the ocean can be summarized by comparing its concentration to the rate it is added or removed from the ocean. The quotient of these is the *residence time* of the element in the ocean. Elements with short residence times (Al, Ti, Fe, Cr) are reactive and form insoluble solids before reaching the ocean or soon after arriving in the ocean. In the latter case, they form authigenic ferromanganese nodules, zeolites, and glauconite. Elements with long residence times (Na, K, Ca, Mg) are characterized by high solubility. Divalent cations have shorter residence times than their monovalent counterparts because they are adsorbed more easily by clay minerals.

In addition, we find that *biological activity* regulates the concentrations of oxygen, carbon dioxide, phosphate, nitrate, nitrite, ammonia, silicate, and possibly vitamins and trace metals dissolved in seawater. These substances are all involved in the metabolism of organisms. Concentrations are also affected by detritus from the disintegration of oceanic organisms. Proteinaceous decay products, for example, associate with zinc, tin, lead, titanium, copper, silver, magnesium, aluminum, chromium, and nickel. Some organisms have the ability to concentrate trace elements by factors as high as 1 million, so significant amounts of these elements may exist in live organisms as well.

Another regulatory role is played by *bacteria,* which fix the concentration of substances involved in biochemical processes in the ocean. In particular, sulfate-reducing bacteria affect the amounts of sulfate and sulfide in oxygen-deficient portions of the ocean. Because they break down organic debris, other bacteria are instrumental in regulating the amount and species of nutrients in seawater.

Physical processes in the ocean tend to regulate the amount of water in seawater rather than the amount of dissolved substances. The formation and melting of ice, evaporation, and precipitation change the absolute concentrations of dissolved substances by dilution or removal of water, but the amounts of dissolved materials remain in the same proportions.

CHEMICAL HISTORY OF THE OCEAN

As conditions on the earth changed during the course of geologic time, there must have been corresponding changes in the chemical composition of the ocean. These chemical adjustments would have stemmed from the regulatory processes that we have just discussed. Much indirect evidence has been obtained by studying ancient sedimentary rock strata on the continents. These continental rocks have weathered under the attack of water containing dissolved atmospheric gases. As a result, most of the major cations have found their way to the ocean. Calcium reached saturation quite early, judging from ancient calcium carbonate and calcium sulfate deposits existing on land today. Potassium accumulated in appreciable quantities as early as 1.2 billion years ago, because sediments of that age contain glauconite, a potassium rich clay-like mineral. An abundance of magnesium caused the formation of strata of dolomite, a calcium magnesium carbonate mineral. The earliest water-laid sediments appear to be about 3 billion years old.

Although the first form of life must have been anaerobic, an organism capable of *photosynthesis* eventually evolved. As photosynthesis progressed, an excess of oxygen was produced that gradually escaped to the atmosphere. The amount of atmospheric and dissolved carbon dioxide decreased, however, by incorporation into organic matter. The hydrogen ion concentration (acidity) of seawater was established and held constant by the *buffering* action of the dissolved silica-clay mineral system and the dissolved carbon dioxide-solid carbonate system (see Section 4-4). As oxygen accumulated, the ocean and terrestrial portions of the earth changed from the primordial reducing condition to an oxidizing one. Only in locations removed from contact with air or oxygenated water do reducing conditions persist today. Such a location would be in the sediments on the ocean floor. Not only are the bulk of bottom sediments isolated from contact with oxygenated seawater, but organic matter incorporated in these sediments decomposes and consumes whatever oxygen may have become trapped in interstitial water when the sediment formed at the bottom. The oxidizing state at the interface of sediment and water has led to the formation of ferromanganese precipitates. These and organic detritus must have began adsorbing trace metals soon after an excess of oxygen accumulated in the ocean and atmosphere.

Considerable geological evidence leads us to conclude that the volume of seawater in the ocean has been increasing throughout the earth's history. The chemical composition of the world ocean, however, became very nearly what it is at present shortly after the advent of photosynthesizing organisms—that is, about 1.5 billion years ago. There have been small-scale changes associated with the evolution of carbonate-secreting organisms and with temperature fluctuations in the Tertiary and Quaternary periods. Increases in the temperature, for instance, would have caused a decreased solubility of dissolved gases and concomitant decrease in biological activity. Under such conditions, nutrient concentrations increase and carbonates precipitate. Nevertheless, except for the possible appearance of some new substance capable of *ion exchange* or *chemisorption*, the effects on regulatory processes in the world ocean today are limited to changes in the rates of inorganic and biochemical reactions.

4.3   QUANTITY OF MATERIAL DISSOLVED IN THE WORLD OCEAN

If we assume that the amount of dissolved material in the ocean remains unchanged even today, we must also assume that the amount of dissolved material entering the sea each year is equivalent to the amount lost in sediment. Actually, the assumption cannot be validated, because scientists have measured the composition of the oceans for only a few decades. If no ions were removed from the sea, the ions added by rivers in 100 years would increase the concentration of substances dissolved in the world ocean by only 0.0005 per cent. Rivers transport approximately 2.5×10^{15} g of dissolved material to the world ocean every year. This amount is minor compared to the amount of dissolved material (5×10^{22} g) already present in the world ocean (20,000,000 times as much as is added every year). In any case, the rate of change is so small that the assumption of a steady state is reasonable.

Every kilogram of seawater in the open ocean contains about 35 grams of ions. In oceanography, this concentration is expressed as parts per thousand and can be written 35 ‰. The measure of the concentration of total ions in seawater is called the *salinity*. A salinity of 35 parts per thousand is an average value for water in the open ocean, although it may actually range from about 33 to 37 parts per thousand.

In nearshore coastal regions, bays, and especially in river estuaries, salinity is highly variable. Near river mouths, the salinity of surface water may vary from nearly zero to 34 parts per thousand, but low values generally prevail. These variations are also seasonal and depend on river conditions. During summer months, when rivers carry less water, nearshore salinities may be close to 34 parts per thousand. During the winter rainy season or spring thaw, the seawater in the nearshore region frequently becomes diluted by floodwater.

Exactly the opposite conditions prevail in hot, dry regions of the world where excess evaporation causes salinities to be higher than average. Salinity values as high as 40 parts per thousand may be found in places like the Mediterranean and Red seas.

THE RULE OF CONSTANT PROPORTIONS

Despite the range in salinity in seawater, the relative proportions of the major ions in the open ocean vary by only negligible amounts. This observation was reported as early as 1819. In 1865, Forchammer noted that the proportions held true in hundreds of samples of surface waters.

Later, Dittmar analyzed 77 samples of water collected from various depths in most of the oceans by the Challenger Expedition (1872–1876). His findings verified the theory that has come to be called the *rule of constant proportions:*

> Regardless of how the salinity may vary from place to place, the ratios between the amount of the major ions in the water of the open ocean are constant.

This rule applies not only to lateral variations within the open ocean, but also to variations with depth.

The rule of constant proportions illustrates two important points. First, the basins of the world ocean are interconnected, and water is exchanged from one basin to another. Thus, the sea is mixed well enough to eliminate differences in composition caused by substances brought to the sea by rivers. Apparently, this elimination happens within the thousand or so years that it takes to mix the ocean completely. Secondly, the rule illustrates that the salinity of the water in the open ocean changes by adding or subtracting water, not dissolved substances. If ions were added or removed, we would expect their relative proportions to change. The only processes that can maintain constant proportions of ions are: evaporation, precipitation, and freezing or thawing of seawater in the open ocean. It is true that along the coastlines ions are added in abundance, and so it is also true that the rule of constant proportions does not hold in these regions.

Nor does this rule hold for the nutrient ions or trace elements. Since the nutrients are utilized by plants, their concentration varies tremendously according to the seasonal life activity of the plants. This activity utilizes nutrient ions near the surface where the plants grow. When organisms die, they sink, decompose, and thus release nutrients to deeper water. Similarly, the concentrations of trace elements do not follow the rule of constant proportion, because their distribution is governed largely by organic activity.

Nevertheless, the rule of constant proportions makes it possible to calculate the salinity of seawater without resorting to determining concentrations of each of the major ions in seawater, which is a time-consuming and tedious task. If one of the major ions is determined, the concentrations of the other seven can be calculated by use of the rule of constant proportions. Scientists usually determine the chloride ion concentration, because it is the easiest to measure chemically.

The usual chemical method of determining the chloride ion concentration also measures the concentrations of bromide, iodide, and several trace anions, so a property called the *chlorinity* can be used in lieu of chloride ion concentration. Chlorinity is defined by the amount of silver required to remove all halogens from 0.3285 kg of a seawater sample.

The salinity or total weight of solids dissolved in seawater can be calculated once the chlorinity is determined.* The salinity of seawater equals the concentration of major ions, trace-element ions, and nutrient ions. The concentration of major ions in a seawater sample is defined as the chlorinity divided by 0.55, or equal to about 1.8 times the chlorinity, or S $\%_0$ = 1.8 Cl $\%_0$. This relationship is an average one that holds fairly well over the entire open ocean.

4.4   DISSOLVED GASES IN THE OCEAN

The most abundant gases in the ocean are oxygen (O_2), nitrogen (N_2), and carbon dioxide (CO_2). These gases are dissolved in seawater and are not in chemical combination with any of the materials composing seawater. Likewise, hydrogen and oxygen chemically combined as water molecules are not to be considered gases dissolved in seawater. The distinction between combined gas and dissolved gas can be made biologically; that is, fish use dissolved oxygen to breathe but cannot utilize oxygen from water molecules.

Gases are dissolved from the atmosphere by exchange across the sea surface. Air of the atmosphere is composed of a mixture of 78 per cent nitrogen, 21 per cent oxygen, 0.9 per cent argon, 0.03 per cent carbon dioxide, water vapor in variable quantities, and small amounts of the noble gases (neon, krypton, helium, radon, argon, and xenon). The gases dissolved at the sea surface are distributed throughout the world ocean by mixing, or *advection*, and diffusion. Concentrations are modified further by biological activity, particularly by plants and certain bacteria.

In nature, gases dissolve in water until saturation is reached, given sufficient time and mixing. The volume of gas that saturates a given volume of seawater is different for each gas and depends upon temperature, pressure, and salinity. An increase in pressure, or a decrease in salinity, or a decrease in temperature causes an increase in gas solubility.

There are small additions of dissolved gases from sources other than the atmosphere. Some nitrogen, for instance, is formed near the bottom by bacterial action on dissolved nitrate ions. Argon is a product of the radioactive decay of potassium; radon and helium are products of the radioactive decay of uranium and related elements.

OXYGEN

Perhaps the most important dissolved gas in seawater is oxygen. Animals require oxygen for respiration. Plants release oxygen as a

---

* In this case, the assumption is made that bromide is replaced by an equivalent amount of chloride, that carbonate is converted to an oxide, and that all organic matter is destroyed.

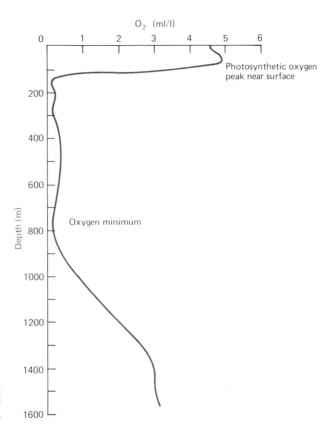

O_2 (ml/l)

Photosynthetic oxygen
peak near surface

Oxygen minimum

Depth (m)

FIGURE 4-3

Typical profile of dissolved oxygen
in the sea (from Eastern Tropical
Pacific Ocean).

by-product of photosynthesis and utilize it during respiration. The decomposition of organic material in the ocean is dependent upon oxygen
concentration. Consequently, the amount of oxygen dissolved in seawater
depends not only on mixing but also upon the type and degree of biological activity.

The amount of oxygen dissolved in the sea varies from zero to a
maximum of about 9 milliliters per liter of seawater. At the surface of the
sea, the water is more or less saturated with oxygen because of exchange
across the surface and plant activity. In fact, when photosynthesis is at a
maximum, it can cause *supersaturation* of seawater. Values of oxygen
concentration in the sea often are expressed as a percentage of saturation
to reflect how closely the theoretical saturation value has been attained.
Oxygen in water found at depth is expressed as *apparent oxygen utilization*
(AOU). These values indicate the difference between the observed concentration and the theoretical saturation concentration; they also reflect
the amount of change in oxygen content since the water sample had last
been at the sea surface.

The vertical distribution of dissolved oxygen is illustrated in Fig.
4-3. At the sea surface, the oxygen content is maintained by mixing surface water and air bubbles and by photosynthetic production. Mixing is

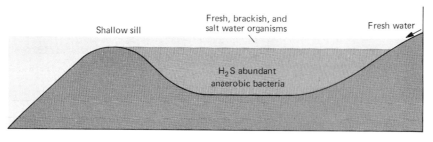

	Surface layer (upper 200 m)	Deep water (200-bottom)
Temp.	15°C(ave.)	8.5°C
Salinity	18‰	22‰
Oxygen	6 ml/l	0 ml/l

FIGURE 4-4

Schematic representation of the Black Sea. This is a famous example of a stagnant basin with a shallow layer of oxygenated surface water overlying a great volume of oxygen deficient water.

effective to a depth of several tens of meters. Below this depth, however, animal respiration and organic decomposition remove oxygen. Dissolved oxygen decreases with depth to a minimum value between about 700 to 1000 meters. Then oxygen values increase slightly with depth, because bottom water, having been at the surface more recently than water at intermediate depth, still contains appreciable dissolved oxygen. The oxygen minimum zone is usually absent in areas of major convergences where surface water is introduced into intermediate depths. (See the chart of ocean currents in Fig. 7-1.)

In places where vertical mixing is restricted by unique hydrographic and geographic conditions, animal respiration and especially organic decomposition remove all the dissolved oxygen from the bottom water. When these conditions occur, organisms called sulphate-reducing bacteria thrive. These bacteria convert sulphate ions (SO_4^{--}) to hydrogen sulphide (H_2S), a foul-smelling and poisonous gas. An example of this situation occurs in the Black Sea and is illustrated in Fig. 4-4. The Black Sea is characterized by a restricted entrance to the ocean, a very large freshwater inflow, and little vertical mixing. Because the deep water cannot be replenished, dissolved oxygen is depleted, hydrogen sulphide is present, and no life except sulphate-reducing bacteria exists below a depth of about 200 meters.

OXYGEN ISOTOPES. The *stable isotope* of oxygen, O^{18}, has a larger mass and is more sluggish in its movement than the more common isotope O^{16}. Therefore, it is *fractionated* during biochemical reactions, so organic matter has a O^{18}/O^{16} ratio that is lower than that in seawater. At elevated temperatures, the heavier isotope is less sluggish, so O^{18}/O^{16} ratios tend to increase with temperature. This effect has been used to ascertain the former temperature of the sea by examining the O^{18}/O^{16} ratio in the remains of ancient marine organisms. Heavy isotopes of other elements in seawater behave similarily, but most of them are not studied because they are either less abundant or do not enter into biochemical reactions.

NITROGEN

Nitrogen is a relatively inert (nonreactive) gas that dissolves in the ocean in concentrations between 8 and 15 ml per liter. These are the concentrations expected in seawater saturated with nitrogen from the atmosphere. The distribution of dissolved nitrogen in the ocean is governed by its solubility (and, hence, temperature), by salinity, and by the mixing of currents. The elaborate cycle of nitrogen through the sea and its organisms is shown in Fig. 4-5. Certain algae living in tropical and subtropical water are capable of incorporating (*fixing*) dissolved atmospheric nitrogen as protoplasm. Where oxygen is present in extremely low concentrations (below 0.15 ml per liter), bacteria convert nitrate ions to molecular nitrogen in a process called *denitrification* (see Chapter 13). Fixation and denitrification probably do not change the quantity of nitrogen dissolved in the nitrogen-saturated ocean. However, denitrification in the ocean must be appreciable. If the amount of fixed nitrogen (nitrate ions and organic debris) lost to sediments were not somehow returned to

FIGURE 4-5

The nitrogen cycle in the sea.

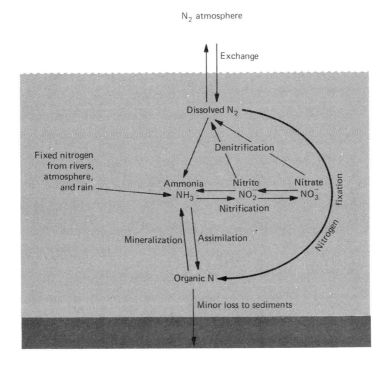

the atmosphere (presumably via denitrification), molecular nitrogen would have been removed from the atmosphere in about 4×10^8 years from the start of life in the ocean.

CARBON DIOXIDE

Dissolved carbon dioxide is exceedingly important in biological processes in the ocean. Plants and animals release carbon dioxide during respiration, and plants use it in photosynthesis. In seawater, a sensitive equilibrium exists between dissolved carbon dioxide, the hydrogen ion, carbonate ion, bicarbonate ion, and solid calcium carbonate in bottom sediments. Biological activity does not significantly change the amount of carbon dioxide dissolved in seawater. If an excess is produced, calcium carbonate dissolves; if a deficiency exists, calcium carbonate will precipitate. These reactions—called the carbonate buffer system—cause the concentration of dissolved carbon dioxide to remain almost constant, between 45 and 54 ml per liter.

Atmospheric carbon dioxide dissolves in ocean water according to its solubility as determined by temperature and salinity. Since the ocean contains vast amounts of dissolved CO_2, carbonate ion, protoplasm, and solid carbonate, it regulates the amount of carbon dioxide in the atmosphere. In turn, carbon dioxide helps to regulate the temperature of the ocean, because carbon dioxide in the atmosphere passes solar radiation energy but absorbs much of the thermal (infrared) energy emitted by the earth. Much of the heat trapped by atmospheric carbon dioxide becomes stored in the world ocean. In due course, an equilibrium involving carbon dioxide in the ocean and atmosphere and the mean temperature has been established. The equilibrium also regulates the amount of water vapor in the atmosphere and hence evaporation and precipitation. It

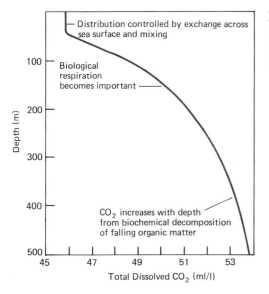

FIGURE 4-6

Generalized profile of the distribution of dissolved CO_2 in the sea off Southern California.

appears that this equilibrium has been upset in the past and the resulting temperature shift has been linked with the glaciation in the Pleistocene epoch.

Like any other dissolved substance, carbon dioxide is distributed in the ocean by advection and diffusion. However, its distribution is also affected by its exchange across the ocean surface and bottom and by its involvement in biochemical reactions. (Fig. 4-6).

NOBLE GASES

Noble gases are those gases that are inactive, especially toward oxygen. The noble gases: helium, neon, argon, krypton, xenon, and radon, dissolve in seawater from the atmosphere just as do the major atmospheric gases. However, once dissolved, they become mixed away from the sea surface. They are distributed only by diffusion and advection, because— being relatively inert—they do not enter into biochemical reactions. Their concentration in seawater reflects the temperature and salinity possessed by that water when they dissolved at the sea surface.

4.5 ANALYTICAL TECHNIQUES OF CHEMICAL OCEANOGRAPHY

In order for a chemical determination to be useful to a chemical oceanographer, it must fulfill several requirements. First, it must measure accurately the concentration of a particular substance even though present in minutely small quantities. Secondly, the mechanics of the determination must be rugged enough to go to sea. All of the equipment must be able to function in a salt-air environment on board a ship that is rolling and pitching under severe weather conditions. Finally, the determination should be rapid. The chemist should be able to keep up with the samples as they are being taken.

Seawater is analyzed for trace-element ions with specialized microchemical techniques adapted from biochemistry and clinical chemistry. Many of these determinations cannot be adopted as standard procedure, because they cannot meet the last two requirements just listed.

The concentrations of nutrient ions in seawater are determined by standard chemical techniques that have been adapted for use at sea. These determinations are made at the time the samples are recovered, or the water is frozen for shipment to a laboratory ashore.

The concentration of the major elements in seawater are usually calculated from the measured concentration of the halide ions (chlorinity) by invoking the rule of constant proportions. This rule permits deriving fixed ratios between each major ion and the chlorinity of seawater. The classic method of measuring the halide ion concentration is by *titration* with a silver salt. In practice, a measured volume of seawater is titrated, and the chlorinity is calculated by referring to tables that convert the weight of silver ions to chlorinity values.

The chemical method of determination of chlorinity, and hence salinity, has been largely supplanted by electrical methods. The conductometric method is based upon the phenomena that seawater conducts electricity in proportion to the amount of ions contained (provided temperature is held constant). An instrument called a *salinity bridge* is used to measure the electrical conductivity of a sample of seawater under controlled temperature (see Chapter 15). From the conductivity, the total of the major ions (salinity) is calculated. A device called a *salinometer* exploits the relation between magnetic susceptibility of seawater and its salinity. A sample of seawater is introduced into the center of a coil, and the measured change in inductance is converted to values of salinity.

READING LIST

HARVEY, H.W., *The Chemistry and Fertility of Sea Water*. Cambridge: Cambridge University Press, 1955. 224p.

RILEY, J.P., AND G. SKIRROW, *Chemical Oceanography*, Vols. I, II. New York: Academic Press, 1965. 712p.

ONE
TWO
THREE
FOUR
FIVE
SIX
SEVEN
EIGHT
NINE
TEN
ELEVEN
TWELVE
THIRTEEN
FOURTEEN
FIFTEEN

important
physical properties
of seawater

The physical nature of the world ocean is determined largely by the physical properties of seawater. In particular, the forces that cause circulation of water in the ocean arise from changes in the physical properties of that water. Temperature and salinity directly affect the density, buoyancy, and stability of seawater and, consequently, the motion of water in the ocean basins. The physical properties of seawater also strongly influence the behavior of heat and light in the ocean, thereby controlling thermal and radiant energy in the sea. Because so many oceanic processes are affected by the characteristics of seawater, we must examine each property in some detail.

5.1   PRESSURE

Pressure in a fluid such as water acts in all directions at any point in the fluid. A fluid flows from a region of high pressure to a region of low pressure; and the greater the pressure differential, the greater the speed of flow. Pressure has virtually no effects upon the volume of water because water is almost incompressible. However, the volume of a compressible fluid (a gas) depends upon pressure.

73

In the ocean; *hydrostatic pressure*, which is the pressure arising from the weight of overlying water, is described by the hydrostatic equation:

$$\text{pressure } (P) = \rho g z \qquad\qquad\qquad\qquad\qquad [5\text{-}1]$$

where ρ = the density of seawater (approximately 1.03 grams per cubic cm), g = 980 cm per sec², and z = the depth below the surface of the ocean in centimeters. Calculations show that hydrostatic pressure in the ocean increases by one atmosphere for each 10 meters of depth below sea level. Therefore, in the deepest part of the ocean (about 10,000 meters), the pressure is about 1,000 atmospheres. It is necessary to consider such excessive pressures in oceanographic work. Divers must take precautions, because pressure affects gases dissolved in blood. Water-sampling bottles must be opened before they are lowered to great depths in the ocean because their walls are not constructed to withstand the tremendous pressure. Furthermore, watertight instrument housings containing air must be designed to withstand the extreme pressures that they will encounter at great depths in the ocean.

5.2   DENSITY OF SEAWATER

The density of seawater varies with changes in temperature, salinity, and, to a minor extent, pressure. Most changes in density occur at the sea surface where solar heating, cooling, precipitation, and evaporation cause the greatest fluctuations in temperature and salinity.

As pressure increases, the density of seawater increases; however, this density increase is slight because seawater is practically incompressible. It is only in the greatest ocean depths that pressure effects are measurable, and even there the changes are small.

When ions are added to a fixed volume of water, its mass increases. In just this way, an increase in salinity produces an increase in the density of seawater (Fig. 5-1). Salinity changes also affect the temperature at which seawater freezes. As water increases in salinity, it must be colder for it to freeze (Fig. 5-2). Thus, seawater freezes over a range of temperatures rather than at a single temperature. For seawater of average salinity, freezing begins at approximately −2° C. As ice begins to form, however, the remaining brine becomes more saline (sea ice is essentially free of ions) and must reach an even lower temperature to freeze.

The effect of temperature on the density of seawater is more complex than that of pressure or salinity. The density of most seawater (where salinity is greater than 25 parts per thousand) decreases with an increase in temperature (Fig. 5-3), so that cold seawater is more dense than warmer water that has the same salinity. Seawater with salinity less than 25 parts per thousand (brackish) becomes less dense if cooled below the temperature at which it reached maximum density.

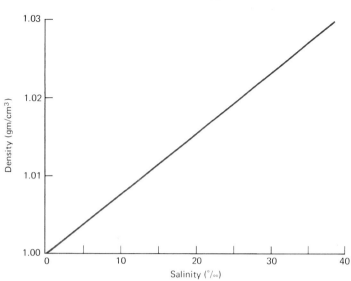

FIGURE 5-1
Curve showing the variation of
density with salinity for water
of 10° C.

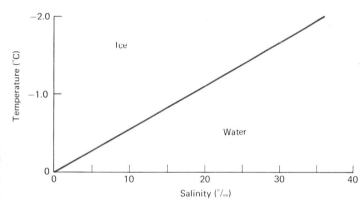

FIGURE 5-2
Curve showing the variation of
freezing point of water as a
function of salinity.

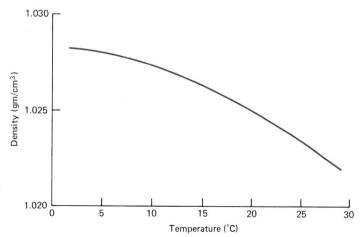

FIGURE 5-3
Variation of density of sea water
(35 ‰) with temperature.

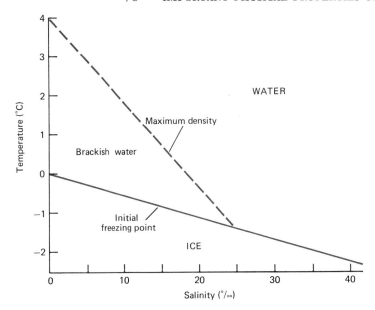

FIGURE 5-4

Effect of salinity on the temperature of maximum density and the initial freezing point of sea water.

The combined effects of temperature and salinity on the maximum density and freezing point of water are demonstrated in Fig. 5-4. Pure water reaches its maximum density at 4° C, but saline water reaches its maximum density at some lower temperature. Most seawater (i.e., having a salinity greater than 25 parts per thousand) has no maximum density peak, because it freezes before this peak is reached. In other words, the density of seawater increases with decreasing temperature until it freezes.

Several generalizations about density in the world ocean ought to be remembered: (1) Seawater has a density between 1.02 and 1.03; in the ocean, density normally increases with depth. (2) Seawater becomes less dense by either warming or dilution (precipitation or thawing), or both. (3) Seawater becomes more dense when it is cooled or when its salinity is increased (by evaporation or freezing), or both. (4) Pressure has little effect on salinity. (5) Seawater density is more sensitive to temperature fluctuations than to salinity fluctuation.

STABILITY IN THE OCEAN

Stability in the ocean is governed by the depth distribution of density.* Density, in turn, is governed by the distribution of temperature and salinity. Not only is density more sensitive to temperature changes

---

* See the Appendix for a discussion of stability.

TABLE 5-1

STABILITY IN THE OCEAN

	Stability	Instability	Indifferent stability
Density	Increase with depth	Decrease with depth	Uniform
General Cases			
Temperature	Decrease with depth	Increase with depth	Isothermal
Salinity	Increase with depth	Decrease with depth	Isohaline
Special Cases			
Temperature and Salinity	Slight increase with depth of temperature plus large increase with depth of salinity	Slight decrease with depth of temperature plus large decrease with depth of salinity	Change in temperature exactly offset by change in salinity

than salinity changes, but the range of temperature in the ocean is greater than the range of salinity. Therefore, temperature is perhaps the most important factor in determining stability in the ocean.

Several combinations of temperature and salinity can lead to stability, instability, or indifferent stability in the ocean. Table 5-1 shows several examples of such combinations.

Typical density profiles in the open ocean are illustrated in Fig. 5-5. These profiles are largely the result of the temperature distribution. The shape of the salinity profile depends upon geographic location. Areas where the salinity decreases with depth are usually areas where temperature decreases rapidly with depth so that stability is maintained.

The typical density profile describes warm water over cold in a stable configuration. Quite different temperature and salinity conditions produce stability in fiords, estuaries, and landlocked seas. In Puget Sound and the Black Sea, for example, cold fresh runoff from rivers lies over warmer saline water flowing in from the sea. This causes extreme stability and is referred to as the *freshwater lid effect*.

In the open ocean, the shape of the salinity profile is governed by geographic effects on the balance of evaporation and precipitation. In the sub-tropical zones (20°−30° Latitude), evaporation exceeds precipitation, so surface water has the highest salinity in the world ocean. The profile of salinity of this surface water shows a decrease with depth. In high latitudes and near the equator, precipitation exceeds evaporation, so

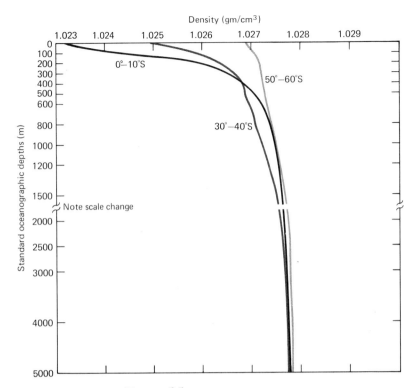

FIGURE 5-5
Typical density profiles from three regions of the world ocean. Note scale distortion of depth to emphasize the surface regions of the column. (Data after Muromtsev, 1958.)

the opposite salinity profile prevails, because the water is diluted near the surface. For the entire world ocean, however, evaporation must exceed precipitation since the amount of water on earth is constant. In other words, total runoff from land plus total precipitation must equal total evaporation.

The evaporation-precipitation balance causes important local effects in estuaries, bays, and lagoons at the periphery of the world ocean. These effects are discussed in detail in the section on inshore oceanography in Chapter 10.

Although an unstable density configuration cannot persist in nature, it does occur in a seasonal manner in middle and high latitudes. When instability occurs, a vertical circulation, or *overturn*, takes place, so that stability is restored. Figure 5-6 shows temperature, salinity, and density profiles taken at the four seasons in the open ocean. During the summer (Fig. 5-6C), heating and evaporation lead to stability. In the fall (Fig. 5-6D), cooling at the surface causes the saline water to become denser than underlying water, so overturn takes place. Mixing by storms tends to destroy the thermocline and enhance the overturn. This process continues throughout the winter (Fig. 5-6A), especially if ice forms at the surface, because the freezing of seawater increases the salinity of the residual brine. By spring (Fig. 5-6B), the warming of surface water and development of a thermocline reestablish a stable density profile that persists through the following summer.

At the continental shelf off Antarctica, ice forms on the surface of the sea every year. Here, therefore, there is an overturn every winter, because the freezing of seawater removes relatively pure water and leaves extremely saline and cold, dense brine that promptly sinks.

In summary, stability occurs when warm water lies over cool water as a result of solar heating at the surface or when fresh water overlies saline water because of river outflow or excessive precipitation. Instability is usually caused by saline water overlying less saline water as a result of evaporation or freezing at the surface. Instability also results when cool water lies over warmer water because of seasonal cooling at the surface. Indifferent stability exists when the temperature and salinity are uniform with depth. In the open ocean, this condition is invariably caused by mixing as a result of overturn, wind action on the sea surface, and currents. In local areas, such as Puget Sound, tidal currents are instrumental in producing mixing that leads ultimately to indifferent stability.

5.3   VISCOSITY

The viscosity of seawater has an important influence on the motions of the world ocean and upon the nature of the forms of life that float or swim in the sea.

Viscosity arises from the internal friction of a fluid. It represents the ease with which molecules in the fluid move past one another or, alternatively, the ease with which mechanical energy can be exchanged between adjacent molecules. It is fluid viscosity that causes a motorboat to stop when the engine has been stopped. The viscosity of a fluid depends upon the chemical nature of its molecules. Syrup is much more viscous than water, so a boat moving in syrup would come to a stop sooner and in less distance than in water. Viscosity is greatly affected by the temperature of the fluid; the viscosity of a fluid doubles if the temperature is decreased by 20° C. On the other hand, salinity has only a minor effect on viscosity. In fact, the viscosity of seawater is about the same as that of fresh water of the same temperature.

5.4   THERMAL PROPERTIES OF SEAWATER

Water has an unusually high heat capacity, latent heat energy of evaporation, and latent heat energy of fusion. These properties (discussed in the Appendix) are significant when considering the thermal behavior of the world ocean.

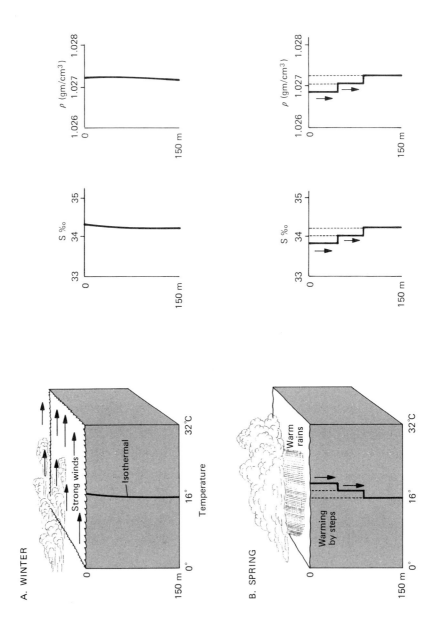

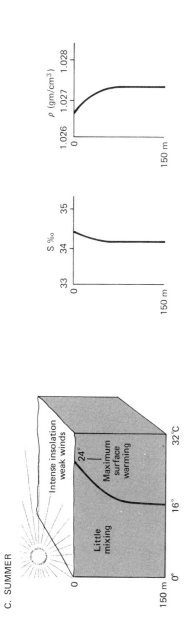

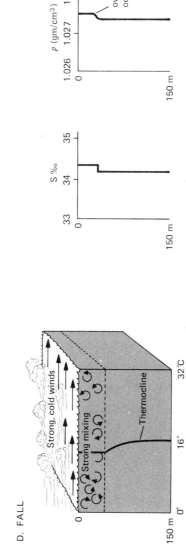

FIGURE 5-6

Schematic representation of seasonal changes in temperature, salinity, and density with depth in a mid-latitude oceanic area. (After Muromtsev, 1958; and Strahler, 1963).

HEAT CAPACITY IN THE OCEAN

The ocean can store a great amount of solar thermal energy with relatively small changes in temperature because of its high heat capacity. A comparison of the ranges of temperature observed on land and in the ocean demonstrate this phenomenon quite readily. On land, a range of about 146° C has been observed between Libya (58° C) and Antarctica (−88° C). In the ocean, the range is only 38 degrees (°C) between the Persian Gulf (36° C) and the Polar Seas (−2° C). These facts show that the ocean acts as a reservoir of heat energy that becomes neither hot in the summer nor cold in the winter. Because of the lower heat capacity of air, the temperature of air varies considerably from day to night, as well as seasonally. In contrast, the temperature of the water of the world ocean remains relatively constant. Therefore, a thermal gradient (or difference in temperature) usually exists between the ocean and the atmosphere. The ocean rapidly exchanges heat energy with the atmosphere. Consequently, near the ocean, air temperatures are thermostatically controlled. For this reason, the climate of oceanic islands and coastal lands has less temperature variation than inland areas.

THE LATENT HEAT ENERGY OF EVAPORATION
IN THE OCEAN

Water's high latent heat energy of evaporation causes large amounts of thermal energy to be supplied to the atmosphere when water vapor evaporates from the ocean. The atmosphere is much more mobile than the sea because of the low viscosity of air, so heat energy from the ocean reservoir can be distributed rapidly over the earth's surface. This energy is released when the water vapor condenses as precipitation in cool regions. Hence, coastal climates are milder than inland climates, especially where atmosphere circulation brings maritime air over cool land.

THE LATENT HEAT ENERGY OF FUSION
IN THE OCEAN

The latent heat energy of fusion is high, so great amounts of heat energy must be released from water before it will freeze. Consequently, relatively little ice forms on the surface of the world ocean, except in extremely cold (heat-deficient) polar regions.

5.5   RADIANT ENERGY AND THE OCEAN

The ocean's source of radiant energy is the sun. The sun emits a spectrum of radiation frequencies, of which 50 per cent are infrared, 41 per cent visible light, and 9 per cent ultraviolet, X rays, and gamma rays. The frequencies denote sizes of quanta. As solar radiation penetrates first the atmosphere and then the ocean, quanta that excite the molecules in air or in water are absorbed. Figure 5-7 shows that an infrared photon is absorbed poorly in molecules in air but quite well in water molecules; that ultraviolet photons are absorbed best by air molecules; and that blue and green photons are least absorbed in air and water. Therefore, infrared radiation passes through the atmosphere easily (the atmosphere is transparent to infrared radiation) but is absorbed by water.

FIGURE 5-7

Relative change in solar energy due to absorption as it passes through the Earth's atmosphere and penetrates through one meter of seawater. (A) Energy spectrum of solar radiation reaching the outer atmosphere; (B), Absorption due to the principal absorbing gases in the atmosphere; (C) Energy spectrum of solar radiation reaching the sea surface; (D), Absorption by sea water; (E), Energy spectrum of solar radiation reaching one meter depth. (Adapted in part from Miller, Merrill Books Inc., and Sverdrup, et al., Prentice-Hall Inc.).

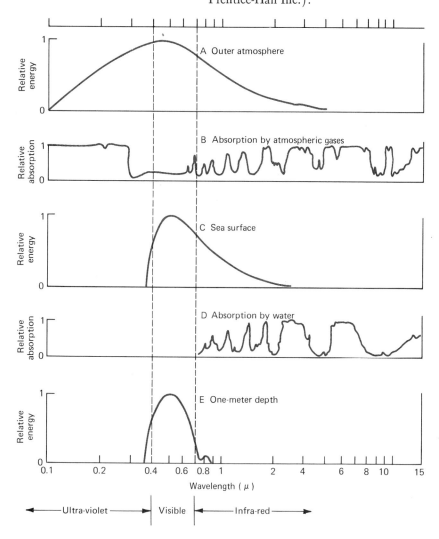

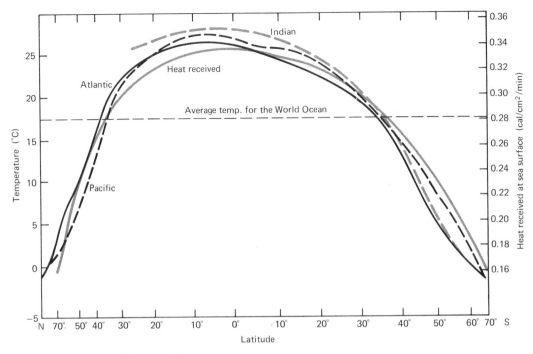

FIGURE 5-8

Mean zonal distributions of surface temperature for the three ocean basins and heat gain through the sea surface as a function of latitude. (Data from Sverdrup, et al., Prentice-Hall Inc.).

Because infrared photons are absorbed into molecules of water the same way that thermal energy quanta are, introducing either infrared photons or thermal energy excites the water molecules. As a result, absorption of infrared photons increases the heat energy and the temperature of the water. Infrared radiation is so strongly absorbed by water that it does not penetrate the ocean deeper than about one meter. Consequently,

FIGURE 5-9

Energy spectrum of solar radiation at the water surface and at various depths. (After Sverdrup, et al., by permission of Prentice-Hall Inc.).

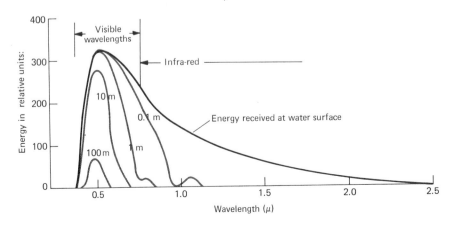

solar heating occurs in only the upper few meters of the world ocean, and heat energy is carried deeper only by conduction and vertical mixing. The general picture of the distribution of temperatures in the ocean is shown in Fig. 5.8.

Blue-green light is absorbed least. It does not cause thermal excitation of molecules, but it is used for plant photosynthesis. It penetrates to about 300 meters in the open ocean and to about 10 meters in harbors and other nearshore areas. Fig. 5-9 shows how radiant energy of several frequencies is absorbed at a particular location in the world ocean. Below about 100 meters, there is virtually no penetration of light, because total absorption of radiant energy occurs below this depth. The zone between the surface and the depth of total absorption is called the *photic*, or lighted, *zone*.

The blue-green color of the ocean is caused by the minimum absorption of blue-green light. Blue-green light can be transmitted into, reflected, and transmitted out of water without being absorbed.

Sometimes, however, color of the ocean is not blue-green. For example, dissolved organic matter of yellowish-brown color gives a green or greenish-brown color to the ocean. There are red tides, caused by dinoflagellates (very small marine organisms) that impart a reddish tint to the ocean. In certain locations, such as off the coast of Norway, microorganisms called *coccolithophore* sometimes concentrate and impart a milky-white color to the ocean. Suspended sediment may cause a muddy coloration near rivers and along shores where wave action is strong.

READING LIST

CHRISTIANSEN, G.S., AND P.H. GARRETT, *Structure and Change*. San Francisco, Calif.: W.H. Freeman and Co., Publishers, 1960. 608p.

Conference on Physical and Chemical Properties of Sea Water, National Academy of Sciences—National Research Council Publication 600. Washington, D.C., 1959. 202p.

KRAUSKOPF, K., AND A. BEISER, The Physical Universe. New York: McGraw-Hill Book Company, 1960. 536p.

ONE
TWO
THREE
FOUR
FIVE
SIX
SEVEN
EIGHT
NINE
TEN
ELEVEN
TWELVE
THIRTEEN
FOURTEEN
FIFTEEN

atmospheric circulation

The physical interaction between the atmosphere and the ocean is critically important to many oceanic phenomenon. In the lower atmosphere, winds are responsible for waves, surface currents, and the exchange of energy in the form of heat and water vapor. Particulate matter ranging from mineral grains, seeds, and spores to carbon and silica spherules that are the by-products of industrial combustion are transported out to sea through atmospheric circulation. An understanding of this circulation, therefore, is a necessary part of the study of oceanography.

At first glance, the pattern of surface winds over the world ocean (Fig. 6-1) appears to be somewhat confused and difficult to interpret. This wind pattern can be understood, however, by first considering winds that might be found on a theoretically nonrotating, water-covered globe, then on a rotating globe on which continents are present.

6.1 WINDS ON AN IDEAL EARTH

We assume for this discussion that the earth is motionless and is covered with water. Surrounding this earth is the atmosphere, whose

86

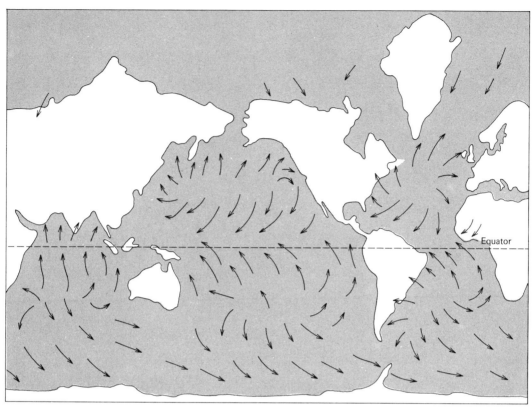

FIGURE 6-1
Surface winds over the World Ocean (average for July).
(After U.S. Navy Hydrographic Office Publication No. 9, 1958)

lower portions are in contact with the sea and are therefore saturated with water vapor. The water and the atmosphere are heated by solar radiation. This heat becomes more intense near the equator. The air in the vicinity of the equator would be heated relative to the surrounding air; its density would decrease; and it would rise. This motion would cause surface air from higher latitudes to flow towards the space left by the rising equatorial air. Actually, the rising air represents a region of relatively low atmospheric pressure. Air, like any fluid, flows toward a zone of low pressure. The equatorial air rises to an altitude determined by its density (i.e., it seeks its own level) and begins to spread laterally toward the poles. As the equatorial air rises, it is cooled, becomes supersaturated and loses moisture by precipitation.

This imaginary system is shown in Fig. 6-2. The equatorial surface regions of the globe are characterized by warm temperatures, low atmospheric pressure, clouds, and rainfall. In the polar regions, the temperatures are low, surface atmospheric pressure is high, and so precipitation will be low.

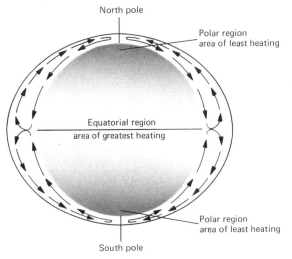

North pole

Polar region
area of least heating

Equatorial region
area of greatest heating

Polar region
area of least heating

South pole

FIGURE 6-2
Ideal atmospheric circulation for a uniform,
nonrotating, nonrevolving earth. (After
U.S. Navy Hydrographic Office,
Publication No. 9, 1958)

6.2   THE ROTATING EARTH

Next, we must consider the effects of the earth's rotation upon the pattern of wind at the surface of the ocean. The fact that the earth is rotating and that we are rotating with it has a startling influence on the way in which we visualize the motion of an object. One example is the movement of the stars as observed from earth. Just because the stars trace a circular path across the sky does not mean that they are actually moving in a circular path. We only see their motion relative to us. If we could free ourselves from the rotation of the earth, the stars would not appear to move in this manner.

It is important to emphasize that the direction of motion as observed by an individual is a relative thing, depending not only on the motion itself but on the vantage point of the observer.

Newton's second law of motion:

$$F = Ma \tag{6-1}$$

states, in part, that a force acting on an object causes that body to accelerate in the direction of the impressed force. It should also be mentioned that when viewed from a place fixed with respect to the stars, this motion will be rectilinear—that is, in a straight line. A problem arises because we are forced to use a reference position that is on the earth and that rotates with the earth. As a result, observations are difficult because an object moving in a straight line (with respect to the stars) past the earth would appear to an observer on earth to follow a curved path. You can verify this by imagining yourself to be riding a carousel, while a person standing at the center is playing catch with another person standing on the ground at some distance away from the carousel. If the ball is thrown at the instant everyone is aligned in a straight line, it would appear as if you would be able to intercept the ball. However, in the time it takes for the ball to reach your position, you have moved in such a way that the ball appears to pass your side. From your rotating vantage point, you would testify that the ball was following a curved path, whereas the person standing on the ground to catch the ball would testify as to its rectilinear motion. Furthermore, you would conclude that a force had to be acting on the ball to account for its continual change in direction with respect to your position. If the carousel had a counterclockwise rotation, the apparent curvature is to the right. A clockwise rotation would cause a curvature to the left.

Our vantage point on the earth is a rotating one. Thus, if we want to describe the motion of an object on the earth by using Newton's second law, his equation of motion must be adjusted to describe the apparent curvature of its path. We make the adjustment by stating that a curvature-producing force is acting on the object, because the only way that a moving object deviates from rectilinear motion is when a force is applied to it. So, to the equation of motion, we add an apparent, or fictitious, "force" that acts 90 degrees to the direction of motion and accounts for the observed deviation. This fictitious force is called the Coriolis effect, and it adapts Newton's second law to observation of motion made on the rotating earth. Mathematically, the Coriolis effect is stated as follows:

$$F_c = 2\omega \sin \phi \, v \qquad\qquad (6\text{-}2)$$

where F_e = Coriolis force per unit mass, ω = angular rotation rate of the earth (a constant), ϕ = degrees latitude, and v = horizontal velocity of the moving body.

The following statements can also be made about the Coriolis effect:

1. It is an apparent force. It cannot do work or initiate motion (when $v = o$, $F_c = o$). It can only alter motion in our eyes.

2. A moving object deviates to its right in the northern hemisphere and to its left in the southern hemisphere.

3. For a given velocity, the magnitude of this force increases with latitude, being zero at the equator ($\sin 0° = 0$) and maximum at the poles.

4. At any given latitude (except at the equator), the greater the velocity, the greater the force.

Note that this effect is only important to the equation of motion when the motion involved is of a geographical scale. In most scientific work and laboratory experiments, Coriolis effects (apparent curvature of motion) are insignificant. Nevertheless, marine and atmospheric scientists who are dealing with movements on a worldwide scale must include the curvature of motion in their equations.

Applying this concept to our rotating earth, we find that as the equatorial air rises and begins to move toward the poles, it is deflected to the right in the northern hemisphere and to the left in the southern hemisphere. This fact is called Ferrel's law. As a result, there is a system of high-altitude westerly winds at the mid-latitudes (Fig. 6-3). Since the northward (or southern) component of flow has been decreased, air tends to accumulate, causing high-pressure regions to occur between 20 and 30 degrees latitude. These subtropical high-pressure belts modify the circulation within the lower atmosphere in such a way that air sinks at about 30 degrees latitude and flows toward both the equator and the poles. The

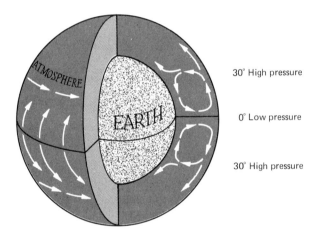

30° High pressure

0° Low pressure

30° High pressure

FIGURE 6-3
Air rising from the equator spreads toward the poles, is deflected eastward, and accumulates at 30° latitude to form the subtropical high-pressure belts. (After Strahler, 1963; and U.S. Hydrographic Office Publication No. 9, 1958).

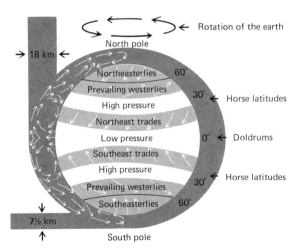

FIGURE 6-4
Simplified diagram of the general circulation of the atmosphere (U.S. Navy Hydrographic Office Publication No. 9, 1958).

air that sinks and moves toward the equator is subsequently warmed and rises, thus completing a tropical circulation system (Fig. 6-3). The surface air moving in a poleward direction from 30 degrees latitude tends to flow over the more dense polar air existing north of 50 to 60 degrees latitude. The confluence between these two air masses is called the *polar front*. It represents a region of highly changeable weather conditions caused by the interaction of relatively warm and moist subtropical air with cold, dry air from the higher latitudes. The net result is the formation of another cell of moving air that rises at about 60 degrees latitude and sinks at the poles.

Figure 6-4 illustrates the winds that result from differential solar heating on a water-covered, rotating earth. The rotational effects are easily seen by the formation of three circulation cells in each hemisphere (called Hadley cells) and the deflection of surface winds to the right in the northern hemisphere and to the left in the southern hemisphere.

This pattern of surface winds forms six zones or belts that travel obliquely toward the poles or toward the equator rather than due north or south. These belts of winds encircle the earth latitudinally and are named the *trade winds*, the *prevailing westerlies*, and the *polar easterlies*.

It is worth noting the climatic conditions that result from the observed air circulation. In the regions of rising air, the climate is characterized by:

1. low atmospheric pressure (rising low-density air)
2. much rainfall and cloudiness

In the zones of sinking air, there is:

1. high atmospheric pressure (sinking dense air)
2. cool and dry air, leading to excessive evaporation

The regions of convergence or divergence of surface winds and the common name given to these areas are also shown in Fig. 6-4. Generally, the *doldrums* are characterized by calms or light winds; the *horse latitudes*, by light and variable winds; and the polar front region, by strong winds. In fact, sailors used to call winds in this polar area as the "roaring forties" and "furious fifties."

6.3   THE EFFECTS OF CONTINENTS ON ATMOSPHERIC CIRCULATION

The inclusion of continental masses on a water-covered, rotating earth modifies the model of atmospheric circulation in two ways. First, the surface winds tend to form closed, horizontal cells over the ocean. Secondly, the distribution of land between the northern and southern hem-

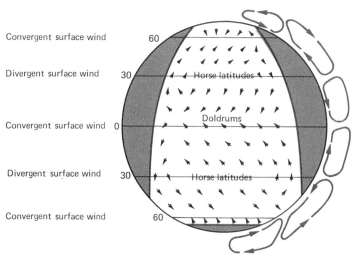

Convergent surface wind

Divergent surface wind

Convergent surface wind

Divergent surface wind

Convergent surface wind

60

30

0

30

60

Horse latitudes

Doldrums

Horse latitudes

FIGURE 6-5
The continental land masses tend to cause the surface winds to form closed cells over the oceans.

ispheres causes the Hadley cells over the ocean to be shifted northward. As a result, the meteorological equator is shifted 5 to 10 degrees north of the geographical equator. These two effects produce the model of atmospheric circulation shown in Fig. 6-5. A comparison between Figs. 6-1 and 6-5 reveals a striking similarity between the actual and theoretical pattern of surface winds.

However, other effects also influence atmospheric circulation. But such effects have only minor influences on the global pattern of atmospheric circulation. Even though they are negligible in the model, they do produce local modifications of the pattern of surface winds.

6.4   SOME LOCAL MODIFICATIONS

Because of the low heat capacity of rock, temperature ranges on continental masses are much greater than that of surface water in the ocean (see Appendix). During the summer, the land mass becomes much warmer than the adjacent ocean. Consequently, the overlying air is heated, then rises, and produces the circulation pattern illustrated in Fig. 6-6. Moist air from over the ocean moves onto the continent to replace the air rising there. The rising air is cooled, becoming supersaturated, thus causing precipitation. During the winter, the land mass is cooler than the adjacent ocean. Air over the land mass is cooled and sinks. This sinking air is relatively dry; and, as it pushes out over the ocean, it increases the evaporation of surface water.

In India, where these effects are pronounced, they are called *monsoons*. The summer monsoon prevails from April to September and is characterized by southwest winds and tremendous rainfall in coastal regions (during the month of July 1861, 366 in of rainfall was recorded at Cherrapunji, a village in Assam, India). During the winter monsoon, the climate is hot and dry; and the winds blow from the northeast for the remainder of the year. The Indian Ocean responds to the reversals in the monsoons by a seasonal change in surface currents.

6.5   DIURNAL EFFECTS

In coastal areas, daily, monsoon-like winds are observed. A steady *onshore* breeze is maintained during the day, because the land is warm relative to the ocean. At night, the land is cooler than the ocean, and so a steady *offshore* breeze blows. Throughout the world, this condition is

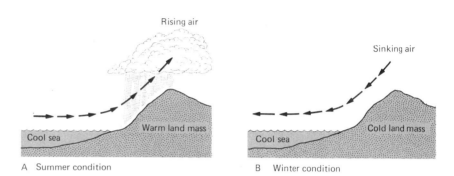

A   Summer condition   B   Winter condition

FIGURE 6-6

Monsoonal wind condition resulting from differential heating between a land mass and adjacent sea. This occurs because of the different heat capacities of rock and sea water. The summer condition is characterized by high rainfall on the windward slopes. The winter condition is characterized by high evaporation over the sea. The west coast of India is a famous example of this condition.

FIGURE 6-7

The effect of a westerly wind (mid-latitude wind belt) on the climatic conditions existing over a continent (winter condition). A section through the 40th parallel across the United States would fit this example.

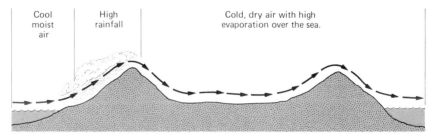

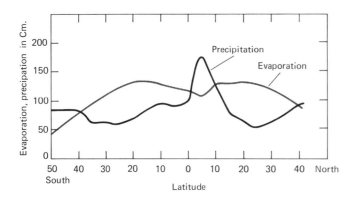

FIGURE 6-8
Latitudinal variation of average
precipitation and evaporation
over the world ocean.
(Data from Sverdrup, Johnson,
Fleming, 1942)

understood by fishermen—they utilize the offshore breezes early in the morning to sail out to the fishing grounds and return in the afternoon with the onshore breezes.

The local modifications of the pattern of surface winds by continental topography explain the difference in climate between the east and west coasts of a continent, especially in the temperate zones in winter. As seen in Fig. 6-7, the west coast is characterized by a cool, wet climate, whereas the east coast suffers from much colder conditions, even though both areas are at the same latitude.

6.6   SURFACE SALINITY AS RELATED TO GLOBAL WIND CHARACTERISTICS

The wind belts (Fig. 6-4) with their associated high and low pressure zones have a controlling influence on the physical properties of the surface waters of the world ocean. This influence is best illustrated by observing the latitudinal variation of the salinity of seawater. We have

FIGURE 6-9
Comparison of average surface salinity and net evaporation
(E-P) for the world ocean as a function of latitude
(After Sverdrup, Johnson, Fleming, 1942)

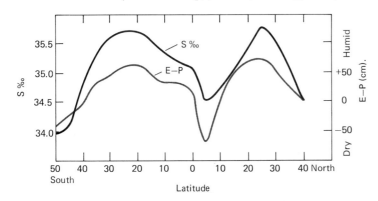

noted that regions of low pressure (at approximately zero and 60 degrees latitude) are associated with high rainfall and that regions of high pressure (30 degrees latitude) have low rainfall. This fact can be illustrated by plotting average values of precipitation and evaporation as a function of latitude (Fig. 6-8). The controlling effect that evaporation-precipitation values have on surface salinity is seen in Fig. 6-9. This figure is a plot of net evaporation (evaporation minus precipitation) and surface salinity versus latitude. The climatic influence on the sea surface is obvious. In the following chapter, other effects of the wind on the sea surface will be discussed.

READING LIST

BOLIN, B., *The Atmosphere and Sea in Motion.* New York: Rockefeller Institute Press in association with Oxford University Press, 1959. 509p.

DONN, W.L., *Meteorology, with Marine Applications* (2nd ed.) New York: McGraw-Hill Book Company, 1951. 465p.

Interaction Between the Atmosphere and the Oceans, National Academy of Sciences–National Research Council Publication 983. Washington, D.C., 1962. 43p.

STEWART, R.W., "The Atmosphere and the Ocean," *Scientific American,* CCXXI, No. 3 (September 1969), 76–86.

ONE
TWO
THREE
FOUR
FIVE
SIX
SEVEN
EIGHT
NINE
TEN
ELEVEN
TWELVE
THIRTEEN
FOURTEEN
FIFTEEN

oceanic

circulation

7.1   A DESCRIPTION OF THE OCEAN CURRENTS

The world ocean is always in motion, and its waters are characterized by many currents (Fig. 7-1). Short-term fluctuations do exist in the direction and magnitude of ocean currents, but, over a period of time, a consistent pattern prevails.

SURFACE CURRENTS

The most conspicuous features in Fig. 7-1 are the large gyres (currents moving in a circle) found in tropical and subtropical regions of each ocean basin. In the northern hemisphere, such gyres move in a clockwise direction, whereas the gyres of the southern oceans rotate counterclockwise. The currents composing the western side of the subtropical gyres are characterized by their increased intensity as they flow to higher latitudes. This westward intensification is best developed in the Gulf Stream and in the Kuroshio Current of the northern Atlantic and Pacific Ocean basins, respectively. These currents attain speeds of approximately 250 centimeters per second. The currents at the western boundaries of the southern oceans also appear to be intensified but to a somewhat lesser degree than in the northern hemisphere.

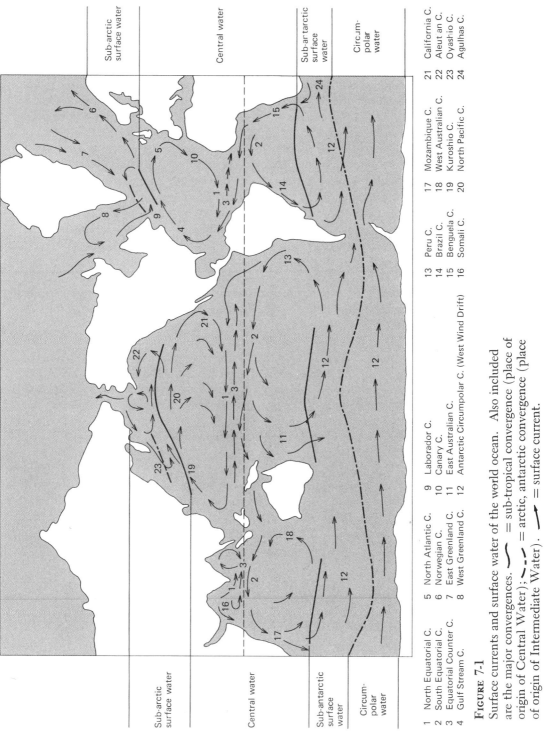

FIGURE 7-1

Surface currents and surface water of the world ocean. Also included are the major convergences. ⌒ = sub-tropical convergence (place of origin of Central Water); ⌒·⌒ = arctic, antarctic convergence (place of origin of Intermediate Water). ⟶ = surface current.

1	North Equatorial C.	5	North Atlantic C.
2	South Equatorial C.	6	Norwegian C.
3	Equatorial Counter C.	7	East Greenland C.
4	Gulf Stream C.	8	West Greenland C.

9	Laborador C.
10	Canary C.
11	East Australian C.
12	Antarctic Circumpolar C. (West Wind Drift)

13	Peru C.
14	Brazil C.
15	Benguela C.
16	Somali C.

17	Mozambique C.
18	West Australian C.
19	Kuroshio C.
20	North Pacific C.

21	California C.
22	Aleut an C.
23	Oyashio C.
24	Agulhas C.

Between the major gyres in the equatorial regions of the ocean basins, countercurrents flow in opposite directions to the adjacent currents. The Pacific Equatorial Countercurrent is well developed over the width of the Pacific basin and attains speeds of 50 cm per sec. The Atlantic Equatorial Countercurrent is generally restricted to the eastern Atlantic Ocean basin. The countercurrent in the Indian Ocean basin is not always present, because it is influenced strongly by the monsoons (see Section 6.4).

The pattern of movement of the colder water of the arctic regions is more complex than that in the lower latitudes. In the northern hemisphere, water movements appear to be strongly influenced by restricting continental boundaries. The North Pacific Ocean basin, which is virtually closed off from the Arctic Ocean basin, contains its own circulation pattern, and there is relatively little exchange to the north. In contrast, the North Atlantic Ocean basin is open to the Arctic Ocean basin, so great volumes of surface water are exchanged between the two basins. In the southern hemisphere, an endless expanse of water surrounds the Antarctic continent. Here, there is little or no obstruction to surface currents. The largest surface flow in the world, the West Wind Drift, flows from west to east around Antarctica.

For convenience, surface water has been named in accordance with its geographical location (see Fig. 7-1). *Central water* occurs within the major gyres of the world ocean. *Subarctic* and *subantarctic water* lies between the *subtropical convergences* and the *arctic* and *antarctic convergences*. The water south of the antarctic convergence is called *circumpolar water*.

DEEP CURRENTS

Water movements are not limited to the surface of the world ocean. Although surface currents are more rapid and more distinctly noticeable to men, they represent only a small percentage of the water within the world ocean. Because of their obscurity, deep currents are difficult to measure. Therefore, little is known about many of the movements within the sea. Nevertheless, motion does exist at all depths.

7.2   THE DYNAMICS OF OCEAN CURRENTS

DRIVING FORCES

The permanent currents of the world ocean originate from an interaction between the atmosphere and the surface of the water. There are two ways in which this interaction occurs. First, a stress is exerted on the sea surface by winds in the lower atmosphere. Secondly, an unstable distribution of density is promoted by freezing or evaporation at the sea surface. In either case, water is put into motion; often both the wind and the climatic effects work together. It is important to understand that even though seawater is in motion at all depths of the world ocean, currents are produced and maintained only at its surface.

There are two types of forces associated with water motion. The first type, called *primary* forces, are those that produce and maintain flow; the second type, the *secondary* forces, arise as a result of motion. In the ocean, the primary forces originate from the interaction of wind and effects of climate on the sea surface. They are:

1. the force of the wind exerted on the sea surface through friction
2. a force related to the relative mass per unit volume of seawater but often expressed in terms of gravity, pressure or density.

The secondary forces are:

1. the Coriolis effect arising from the earth's rotation
2. the forces of friction

In order to analyze the interplay of the forces producing motion of water in the ocean, we should construct a conceptual model in the form of physical equations. All the forces just listed must appear in an equation that describes the motion of water. The equation would state that the forces of gravity, friction, and the Coriolis effect must balance, or else a water particle will accelerate. However, if dynamic equilibrium exists, the sum of these forces will equal zero.

The equations that describe water motion in the ocean are too difficult to solve without simplification. In cases where some of the forces appear to be unimportant, they are ignored and the solutions of simplified equations are compared to the natural conditions to determine if the simplifying assumptions are valid. If the comparison is unfavorable, a reevaluation of the problem must be made. This technique helps an oceanographer understand the forces important in different situations and how various oceanographic phenomena interact.

SURFACE CIRCULATION

The surface layer of water in the ocean is a shallow lens of warm, saline water floating on an immense volume of colder, less saline water. This water is warmed by large amounts of solar radiation in the subtropical regions and becomes relatively saline by evaporation. The boundary between the surface layer and deep water is marked by a sharp *thermocline*. Because of its different temperature and salinity, the surface layer moves in a system of currents that has only a slight physical connection with deep circulation.

Most of the energy driving the surface currents is derived from the wind. Hence, it is necessary to understand just what happens when a steady wind blows over the sea surface and imparts energy to the water.

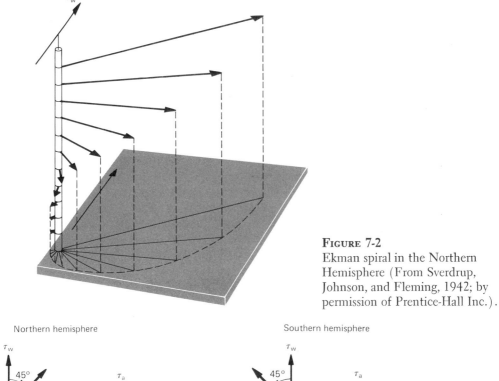

FIGURE 7-2
Ekman spiral in the Northern
Hemisphere (From Sverdrup,
Johnson, and Fleming, 1942; by
permission of Prentice-Hall Inc.).

Northern hemisphere

Southern hemisphere

τ_w

$45°$

τ_a

F_f

F_c

F_f

F_c

τ_w

$45°$

τ_a

F_c

F_f

F_f

F_c

Surface
layer (A)

Lower
layer (B)

Surface
layer (A)

Lower
layer (B)

FIGURE 7-3
Force diagram for a wind driven current. τ_w = wind stress; τ_a = stress exerted
by surface layer (a) on the lower layer (b); F_c = Coriolis force; F_f = force of
friction (coupled with underlying layers). Large arrows represent direction
of water motion.

Observers have noted that the drift of icebergs in the northern seas
deviates 20 to 40 degrees to the right of the wind direction. In 1902,
V.W. Ekman explained this phenomenon by considering the forces produced by a steady wind blowing over the sea surface (frictional wind
stress) and the forces arising from the earth's rotation (Coriolis) and the
force of friction. He assumed a homogeneous sea (no pressure or gravity
forces) and steady state (no acceleration). The results of this famous
work indicated that:

1. Under the influence of a steady wind, surface water will flow 45 degrees to the right of the wind in the northern hemisphere and 45 degrees to the left in the southern hemisphere.
2. At depth, the direction of motion of each water layer deviates to the right relative to that of the overlying water layer, because each lower layer is swept on by the motion of its overlying layer, just as the topmost one is swept on by the wind.
3. The speed of each succeeding layer is less than the one above because of frictional losses as energy is transferred to underlying layers.

These three conclusions can be illustrated by a logarithmic spiral known as the *Ekman spiral* (Fig. 7-2). This spiral also shows that at some depth the direction of the current is oriented 180 degrees to its surface counterpart; however, the magnitude of this deeper current is much less than that at the surface (approximately 1/23 as great). This depth is called the *depth of frictional resistance,* below which the effects of the wind are negligible. A characteristic value for the depth of frictional resistance is 100 m. A force diagram for this model is given in Fig. 7-3.

The most important result of this phenomenon is the average direction in which water is transported. If the direction and magnitude of all of the arrows from Fig. 7-2 are averaged, the resultant vector would be

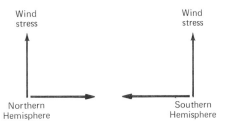

FIGURE 7-4

Direction of the Ekman transport (large arrows) relative to the wind.

FIGURE 7-5

Direction of the surface water as compared to the average direction of the total layer of motion due to wind stress. The paper moves in the direction of the surface water whereas the weighted pole represents the average movement. Large arrow represents the wind direction.

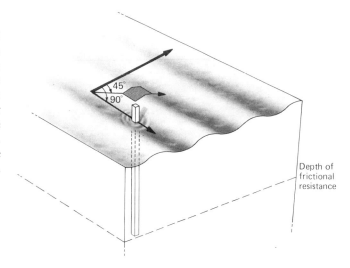

oriented 90 degrees to the right of the wind in the northern hemisphere and 90 degrees to the left in the southern hemisphere. Thus, the mean transport of water (called *Ekman transport*) resulting from a steady wind blowing over the sea surface is oriented 90 degrees to the right of the wind in the northern hemisphere and 90 degrees to the left in the southern hemisphere (Fig. 7-4). To summarize, a piece of paper floating at the surface of the ocean in the northern hemisphere will drift approximately 45 degrees to the right of the wind direction, whereas a wooden rod 100 m long floating vertically will drift 90 degrees to the right (Fig. 7-5).

Comparisons of the Ekman theory with real observations are exceedingly few because of the difficulty in finding situations where all of Ekman's assumptions hold true and no secondary effects are present. Probably the best evidence in support of this theory is the *upwelling* effect due to Ekman transport in the vicinity of a boundary.

When a north wind blows along the west coast of the United States, for example, water is transported 90 degrees to the right of this wind (to the west), thus moving the surface water away from the Washington and Oregon coasts (Fig. 7-6A). Because of the coastal boundary, the water that moves in to take the place of the water removed must come from depth. The effect, therefore, of this particular wind is to bring cold,

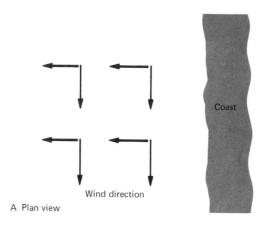

A Plan view

FIGURE 7-6
The process of upwelling, illustrating the initial effects of a north-wind blowing along the west coast of North America. The Ekman Transport (shown by large arrows) is to the west. The small arrows in B represent direction of water motion. The wavy line is the plane around which the motion exists. The symbol ⊙ indicates a wind moving toward the reader. (Modified from Defant, 1961.)

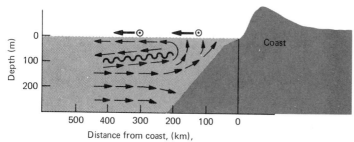

B Profile view

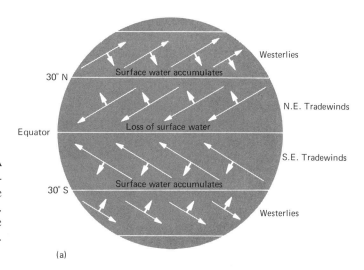

FIGURE 7-7A

Movement of surface water resulting from the Ekman Transport due to the surface winds over the sea. Short arrows represent the direction of Ekman Transport.

30° N   Surface water accumulates   Westerlies

N.E. Tradewinds

Equator   Loss of surface water

S.E. Tradewinds

30° S   Surface water accumulates

Westerlies

(a)

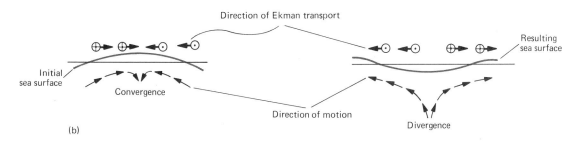

Direction of Ekman transport

Resulting sea surface

Initial sea surface

Convergence

Direction of motion

Divergence

(b)

FIGURE 7-7B

The shape of the sea surface and water motion associated with a convergence and divergence. The symbol ⊙ indicates a wind moving toward the reader and ⊕ away from the reader.

slightly more dense water to the surface along the coast (Fig. 7-6B). This phenomenon is called upwelling. If the wind blew in the opposite direction, the exact opposite effect would occur. The surface water would be "piled up" along the coast and therefore would *sink*. Various combinations of conditions causing upwelling or sinking are possible, depending on the hemisphere, the shape and orientation of the coast, and the wind direction.

It is probably the secondary effects accompanying this phenomenon that make it so noticeable. Not only can this colder, upwelled water affect the coastal climate and, in turn, decrease the enthusiasm of bathers, but it can also cause a redistribution of local fisheries because of the redistributed temperature and salinity conditions.

In addition, the chemical properties associated with the upwelled

103

water cause another important process. This water comes up from depths of 200 to 300 m, so it is often quite rich in plant nutrients (relative to the surface water). Upon reaching the sunlit surface regions, it thus causes increased *phytoplankton* growth.

These illustrations of the importance of Ekman transport should not lead us to believe that a physical boundary is the only type of obstacle that can cause either upwelling or sinking. As illustrated by Fig. 7-7A, a

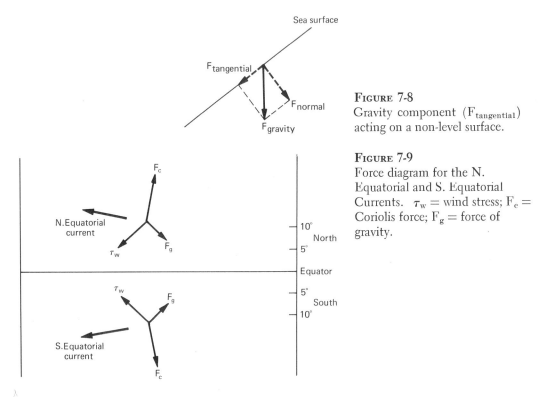

FIGURE 7-8
Gravity component ($F_{tangential}$) acting on a non-level surface.

FIGURE 7-9
Force diagram for the N. Equatorial and S. Equatorial Currents. τ_w = wind stress; F_c = Coriolis force; F_g = force of gravity.

zonal wind condition can cause the same results. Note that a *divergence* is somewhat analogous to an upwelling condition in the open ocean (Fig. 7-7B), and a *convergence* is analogous to the sinking condition (Fig. 7-7B). It is also important to observe the changes in the shape of the sea surface due to convergences and divergences.

The surface currents of the world ocean are a direct result of these zonal winds that blow more or less steadily over the sea surface. Thus, warm, low-density surface water is transported at right angles to the wind direction and accumulates between the wind zones. The best example is the zone between the trade winds and the westerlies (Fig. 7-7A), in which it has been calculated that the accumulated water causes the sea surface to rise approximately 60 centimeters.

Because of the sloping surface of the ocean, there is another force causing the movement of surface water. Besides the force of the wind and the Coriolis effect, which the Ekman theory takes into account, the force of gravity must also be considered. The sea surface is not really a level surface since a component of gravity is present (Fig. 7-8). Thus, a more refined model of the currents in the surface layer of the world ocean should include:

1. the force of the wind
2. the force of gravity
3. the force of the Coriolis effect

These forces can be illustrated by the vector diagrams shown in Fig. 7-9. Note that the gravity component opposes the direction of Ekman transport. This situation is to be expected because it is this transport that causes accumulation of water, whereas the force of gravity acts down the slope (like a ball, which rolls down an incline under the influence of gravity).

FIGURE 7-10
Surface winds over the World Ocean (average for July).
(After U.S. Navy Hydrographic Office Publication No. 9, 1958.)

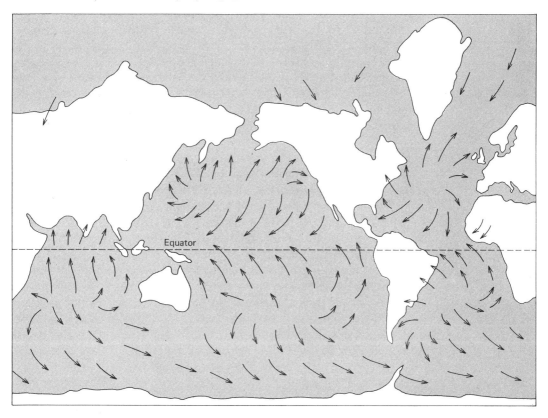

The direction of the current as determined by this diagram is more or less parallel to the wind direction (within 45 degrees). As a rule, such a wind-induced current moves with a speed that is approximately 1.5 per cent of the wind speed (in moderate and high latitudes). The correspondence between the wind and the current is best illustrated on charts showing surface currents and the atmospheric circulation (Figs. 7-1 and 7-10).

One very important current seems to oppose this pattern. It is the Equatorial Countercurrent that is present in all oceans. One simple hypotheses states that this current also is produced and maintained by wind stress, gravity, and the Coriolis effect but in a slightly different association. The asymmetry of the countercurrent about the equator reflects the asymmetry of atmospheric circulation (i.e., it proves the existence of a meteorological equator).

Water driven westward by the trade winds forms the North and South Equatorial currents. These currents pile up water against the east coast of continents. For example, the inclination of this accumulated water in the Atlantic Ocean basin is thought to be approximately 4 centimeters per 1,000 km. Most of the surface water turns poleward after reaching the continents. However, some of it flows back toward the east as a countercurrent, which then flows downward within the calm (called the doldrums) that exists between the trade winds.

Because the development of the countercurrent is determined by the amount of water accumulated, the shape of the continents has an important influence (see Fig. 7-1). The bulge of the northeast coast of South America deflects the Atlantic Equatorial Current northward, so the Atlantic Countercurrent is only of moderate extent. In contrast, the coast of Southeast Asia is shaped like a basin and causes much greater accumulation. Hence, the Pacific Countercurrent is well developed and extends across the entire Pacific Ocean basin. In the Indian Ocean basin, the countercurrent is seasonal because of the influence of the monsoons.

The maximum speed of the countercurrent is approximately 50 centimeters per second, and the volume transport is about 25 million cubic meters per second. The speed of the North and South Equatorial currents is approximately 100 centimeters per second.

Up to this point, we have discussed only the causative forces and the configuration of the surface circulation. In order to complete the picture, a third dimension must be considered—that is, the thickness of the surface layer and the vertical circulation within it.

The lower limit of the thermocline varies between approximately 400 meters (at the equator) and 900 meters (30 degrees latitude). For practical considerations, however, the lower limit of the active surface layer can be considered as approximately 200 to 300 meters. At this depth in the middle and low latitudes, there is a strong pycnocline below which relatively vigorous circulation is lacking (Fig. 7-11). The thickness

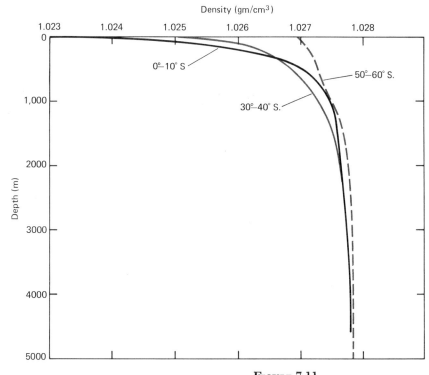

Density (gm/cm³)

FIGURE 7-11
Profiles of average water density for three latitudinal zones in the
Pacific Ocean. (Data from Muromtsev, 1958.)

of the surface layer varies according to the presence of convergences and
divergences produced by circulation. This vertical circulation is caused by
the effects of both winds and climate. For example, the force diagram in
Fig. 7-9 shows a slight poleward component, which, by itself, would cause
a divergence between the North and South Equatorial currents (and likewise a convergence between the equatorial and more northern currents).
Superimposed upon this pattern are the effects of climate on the sea surface; that is, the climate causes density changes in the surface water and
thus changes vertical circulation within the surface layer. Therefore, at
mid-latitudes, under the subtropical highs, excessive evaporation causes the
surface water to sink because of increased density. In contrast, the density
of equatorial water is decreased by the excessive precipitation in the doldrums. Cells of vertically circulating water are produced, and they change
the depth of the surface layer as a result of these convergences and divergences.

A schematic diagram of this circulation is represented in Fig. 7-12.
Remember that the meridional circulation (represented by arrows) is indeed very sluggish as compared to the east-west components. This fact
does not imply, however, that meridional flow is not important. The

107

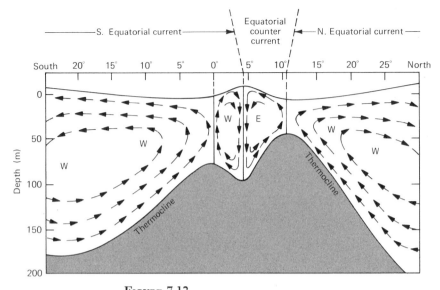

FIGURE 7-12
Schematic representation of the zonal and meridional components of the
surface circulation in the Atlantic Ocean (the topography of the
thermocline is exaggerated in the vertical scale about 1:1 million; that of the
physical sea surface even more); W indicates current toward west; E indicates
current toward east. (After Defant, 1936).

equatorial divergences are important biologically because nutrient-rich
water is welled up into the more sunlit surface regions, thus enhancing
otherwise poor biological growth in the equatorial regions of the world
ocean.

By combining Figs. 7-1, 7-11, and 7-12 conceptually, we can visualize the average shape, physical properties, and currents within the surface
layer. The picture that results differs from the real ocean only with respect
to some details. For example, it does not explain the westward intensification. It has been postulated that the westward intensification of the subtropical currents are a result of:

1. the force of the wind
2. the friction against continental boundaries
3. the change of the Coriolis parameter with latitude

Because these currents in the western part of each major gyre are very
narrow and yet must transport the same quantity of water as the equatorial currents (except for that lost to countercurrents), they must be
deeper and swifter. Within the core of the Gulf Stream, speeds of greater
than 200 cm per sec have been measured. Slower speeds still persist at
depths as great as 1,000 m.

In the Pacific, the Pacific Equatorial Undercurrent moves 125 to
150 cm per sec to the east; it lies at 100 m under the South Equatorial

Current at the equator. This departure from the simple scheme in Fig. 7-12 is not completely understood.

DEEP CIRCULATION

Approximately 90 per cent of the volume of the world ocean is contained below the surface layer, even though the currents flowing in this region are produced and maintained within an area that comprises only about 25 per cent of the earth's sea surface.

In contrast to the surface layer where wind stress is the primary driving force, deep circulation is driven by density differences within the water. These differences are produced and maintained by the climate that exists poleward of the meteorologic and oceanic polar front. (Note the arctic and antarctic convergences in Fig. 7-1.)

Movements within the deep sea exist because water will seek its own level with respect to density. In the polar and subpolar parts of the world

FIGURE 7-13A
Steady-state force diagram for a mass distribution current in the N and S. Hemispheres respectively. F_g = force of gravity; F_c = Coriolis force.

FIGURE 7-13B
In the northern hemisphere particle motion is deflected to the right as it accelerates in response to a sea surface slope. Positions 1, 2, and 3, represent the accelerating phase of the motion. Position 4 represents a steady state. F_g = gravitational force; F_c = Coriolis force; the heavy arrow points to the direction of motion.

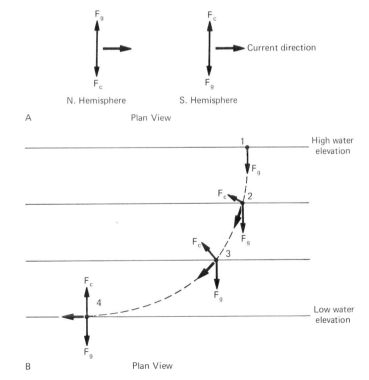

ocean, surface water, cooled strongly by the prevailing climatic conditions or made highly saline by the freezing of sea ice, becomes more dense than the underlying water. As a result, this water will sink to a level at which it will be neutrally buoyant with respect to the surrounding water.

Only two forces are influential in this type of motion, the force of gravity and that of the Coriolis effect. Instability resulting from surface climatic conditions causes water to sink because of its increased density. Since there is motion, the Coriolis effect is also present. The type of current that is produced and maintained by gravity and the Coriolis effect is called a *mass distribution*, or *geostrophic current*. It is depicted by the force diagrams in Fig. 7-13. Note that the steady-state movement is oriented 90 degrees to the gravitational force.

Actually, there are several distinct places in the subpolar and polar parts of the world ocean where sinking takes place. Because these places are at different latitudes and the climate becomes more severe with increasing latitude, the water that sinks within each of these places has different densities and therefore sinks to different depths. As a result, deep ocean water is separated into individual *water masses*, each having independent physical characteristics, movements, and origin. Water that sinks at the subarctic convergence (Fig. 7-1) flows to relatively shallow depths and forms the *intermediate water mass*. As we might expect, the more dense water flowing beneath the intermediate water mass and filling deeper parts of the ocean basins is produced in the higher latitudes where the climate is most severe. Thus, the *deep* and *bottom water masses* of the world ocean are derived mainly from the arctic and antarctic regions of the world.

To summarize, the water filling the world ocean is arranged vertically in layers that originate at the sea surface at higher latitudes and extend continuously to the depths in tropical regions. In this way, a latitudinal arrangement of the water masses at the surface of the world ocean shows the same arrangement as the succession of water masses with depth (Fig. 7-14). The succession is: (1) central water mass (within the surface layer), (2) intermediate water mass, (3) deep water mass, and (4) bottom water mass. The transition between the central water mass and the underlying intermediate water mass is rather abrupt, but the other boundaries are quite diffuse, and each layer grades into its neighbor.

7.3   THE WATER MASSES

Before discussing the individual water masses—their origin, physical characteristics, and movement—we should make some general statements concerning differences in surface water in various regions of the world ocean:

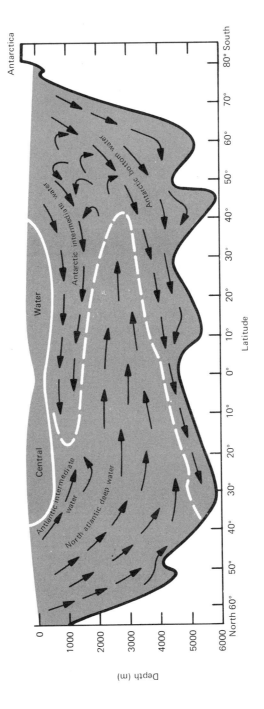

FIGURE 7-14
Schematic diagram of the circulation of the Atlantic Ocean. Arrows indicate the direction of water movement. The boundaries between the water masses is very diffuse.

1. The water in the North Atlantic Ocean basin is characterized by the highest surface salinities found in the open ocean.
2. The North Pacific Ocean basin is characterized by net dilution of surface water.
3. The surface water in the antarctic regions of the world ocean is best described as cold and dilute.

We must consider these generalities in studying the origins of water masses. The water that sinks in a given region possesses (to varying degrees) the surface water characteristics of that region; therefore, when a body of water sinks and flows to another area, its physical characteristics may serve as an indicator of its origin. In this manner, water masses can be traced thousands of kilometers from their origin if we know their temperature and salinity. With time, the characteristics of a given water mass will change, because it mixes with other water masses. However, within limits, this method of tracing water is a valuable tool. Figure 7-15 gives an example of a *T-S diagram* used to "tag" water masses. This type of diagram illustrates the differences among water masses in terms of their temperature and salinity characteristics.

CENTRAL WATER MASS

The name *central water mass* is the general term used to describe the water existing between the subtropical convergences and extending down to the permanent thermocline (Figs. 7-1 and 7-11). In actuality, this water mass can be subdivided into more local units that have about the same density but differ in other physical characteristics, depending on the location. Table 7-1 lists those subdivisions and gives their physical characteristics near the surface and at a depth of 500 meters.

It is clearly shown in Table 7-1 that the North Atlantic surface water is the most saline water in the open ocean. The excessive dilution characteristic of water in the North Pacific Ocean basin is illustrated by the low salinities of the North Pacific central water mass.

To summarize the data given in Table 7-1, the central water mass of the world ocean is characterized by temperatures that range from 7 to 18 degrees (C) and salinities from approximately 34.0 parts per thousand to 36.5 parts per thousand.

INTERMEDIATE WATER MASS

This water mass directly underlies the central water mass and is formed in the regions of the arctic and antarctic convergences. It extends to depths of 1,500 m. Because the intermediate water mass originates in more than one area, a precise temperature and salinity value cannot be

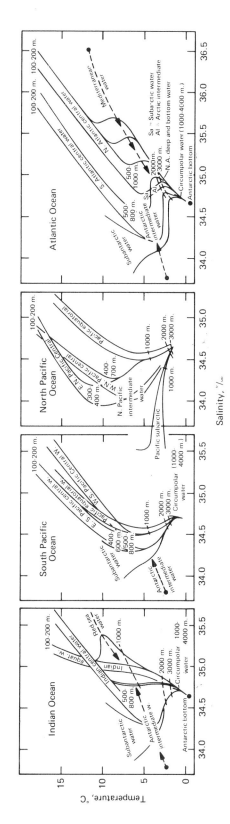

Figure 7-15

Temperature-salinity relations of the principal water masses of the oceans (From Sverdrup et al., 1942, by permission of Prentice-Hall Inc.).

TABLE 7-1

CHARACTERISTICS OF SUBDIVISIONS OF THE CENTRAL WATER MASS
(Data from Sverdrup et al., 1942)

Name	Surface**		500 Meters Depth	
	Temp.	Salinity	Temp.	Salinity
N. Atlantic CWM*	18°C	36.4	7°C	35.1
S. Atlantic CWM	18	35.9	7	34.6
W. So. Pacific CWM	16.5	35.6	7	34.5
Pacific Equatorial CWM	16	35.2	9	34.7
E. So. Pacific CWM	18	35.3	9	34.3
Indian CWM	16	35.6	7.5	34.5
Indian Equatorial CWM	16	35.2	7	35.0
W. No. Pacific CWM	18	34.8	9	34.3
E. No. Pacific CWM	18	34.9	10	34.1

  * C.W.M. refers to central water mass.
** These values actually apply to the upper 100–200 meters.

assigned; instead, it must be characterized by a range of values that vary according to the origin of the water. As in the case of the central water mass, the intermediate water mass can be subdivided according to origin and physical characteristics. The subdivisions are given in Table 7-2.

TABLE 7-2

SUBDIVISIONS OF THE INTERMEDIATE WATER MASS ACCORDING TO ORIGIN*

Name	Temperature	Salinity
N. Atlantic intermediate water mass	3–5°C	34.7–34.9
Antarctic intermediate water mass	2–7	34.1–34.7
N. Pacific intermediate water mass	6–10	34.0–34.1

* Ranges include physical characteristics regardless of location.

Of these subdivisions, the Antarctic intermediate water mass is most extensive. This water originates from the subantarctic surface water that surrounds the earth between the subtropical and antarctic convergences (Fig. 7-1). Because this region is so widespread, the water that sinks mixes to a certain degree and finally becomes the Antarctic intermediate water mass found in all ocean basins. This water flows northward to about 20 degrees north latitude in the Atlantic Ocean basin and approximately 10 degrees south latitude in the Pacific and Indian Ocean basins.

The origin of water masses in the northern basins of the world ocean is restricted by the geometry of the continents and ocean basins in the northern hemisphere (see Fig. 7-1). In fact, the formation of the Arctic intermediate water masses (North Pacific and North Atlantic) is restricted to the western portions of the Pacific and Atlantic basins. In the North Pacific Ocean basin, subarctic water cools and sinks to become the relatively dilute North Pacific intermediate water mass, which has the temperature and salinity given in Table 7-2. The intermediate water mass formed in the North Atlantic ocean basin also originates from subarctic surface water, which has been cooled in the winter. Because, however, of the high evaporation over the North Atlantic Ocean basin, the North Atlantic intermediate water mass is relatively saline (Table 7-2).

There is water in the world ocean that is so homogeneous that it has a single temperature and salinity and thus can be represented by a single point on a T-S (temperature-salinity) diagram. Such water is called a *water type*. It is thought that the water masses of the world ocean are formed by mixing huge amounts of two or more water types. Two important intermediate water types are formed in the Mediterranean and Red seas. The Mediterranean water type leaves the Straits of Gibraltar with a salinity of 38.1 parts per thousand and a temperature of about 13 degrees (C). It flows below the North Atlantic and Antarctic intermediate water masses; and, even though its temperature and salinity change rapidly, it can be traced over most of the Atlantic Ocean basin.

Much less is known of the Red Sea water type. It has a temperature of about 10 degrees (C) and salinity of approximately 35.8 parts per thousand. The water type extends over much of the equatorial and western regions of the Indian Ocean basin.

DEEP WATER MASS

By far the most conspicuous source of the deep water mass lies in the Labrador and Irminger seas of the western North Atlantic Ocean basin (Fig. 7-1). This water results from the mixing of high-salinity water from the Gulf Stream and subarctic surface water. In the winter, the mixture cools and sinks. It flows below the intermediate water mass. As it sluggishly moves southward, it rises over the more dense Antarctic bottom water mass (Fig. 7-14). Thus, the North Atlantic deep water mass extends completely to the floor of the North Atlantic Ocean north of 30 degrees latitude, but it is sandwiched between the intermediate and bottom water masses south of this latitude. The Atlantic deep water mass is characterized by a temperature of 3° C and a relatively high salinity of 34.9 parts per thousand.

No deep water mass is formed in either the Pacific Ocean basin (except for small amounts from the Okhotsk Sea) or Indian Ocean basin.

BOTTOM WATER MASS

The Antarctic bottom water mass contains the most dense water of the world ocean. It is formed off the Antarctic continent in the winter and spreads northward into all three ocean basins. This water mass has been traced as far north as 30 degrees N latitude (Fig. 7-14.)

Freezing plays an important part in the formation of the Antarctic bottom water mass. In winter, vast amounts of water are frozen over the continental shelf of the Weddell Sea. The resulting brine flows down the continental slope and mixes with approximately equal parts of circumpolar surface water to form a water mass with temperature of $-0.5°$ C and salinity of 34.7 parts per thousand.

It should be noted that all of the water masses mix with surrounding water as they flow along. Therefore, they slowly change and lose their identity as they move away from their source. This whole system of oceanic circulation is like a great engine, however, and none of this water becomes stagnant. It is always in motion; it mixes, returns to the surface to be cooled, evaporated, or diluted and sinks again to continue the cycle of oceanic circulation. Studies of the distribution of naturally occurring radioactive carbon in the sea (see Chapter 14) indicate that the mixing cycle of bottom water takes about 1,000 to 1,600 years in the Pacific Ocean basin and about half that time in the Atlantic and Indian Ocean basins. These mixing times (also expressed as *turnover rates* or *residence times*) are verified by corresponding observations using radium (Ra^{226}).

The amount of radiocarbon in a body of water depends upon the age of the water and the degree of mixing with masses of water with different radiocarbon concentrations. Bottom water in the Pacific Ocean basin mixes relatively little, but it has low concentrations of radiocarbon. These facts indicate that the water is rather old. The radiocarbon concentrations at the bottom of the Pacific decrease from south to north in a manner suggesting that the water movement from the Antarctic northward proceeds at about 0.05 centimeters per second.

Surface water has a residence time of 10 to 20 years and is considerably younger than the deep water in the world ocean. These conclusions are based on assumptions concerning the nature of mixing in surface waters and the exchange of radiocarbon between the ocean and atmosphere.

READING LIST

MUNK, W., "The Circulation of the Oceans," *Scientific American*, CXCIII, No. 3 (September 1955), 96–104.

PICKARD, G.L., *Descriptive Physical Oceanography*. New York: Pergamon Press, 1968. 200p.

SVERDRUP, H.U., M.W. JOHNSON, AND R.H. FLEMING, *The Oceans*. Englewood Cliffs, N.J.: Prentice-Hall, Inc., 1942. 1087p.

VON ARX, W.S., *An Introduction to Physical Oceanography*. Reading, Mass.: Addison-Wesley Inc., 1962. 422p.

ONE
TWO
THREE
FOUR
FIVE
SIX
SEVEN
EIGHT
NINE
TEN
ELEVEN
TWELVE
THIRTEEN
FOURTEEN
FIFTEEN

waves

We are all familiar with the ripples that a light breeze can cause on the surface of a fluid. In the sea, any disturbance results in the formation of waves that extend outward from that disturbance. Hence, the waves breaking on a beach may have originated in a storm located thousands of kilometers away. In this chapter, we investigate the origin and nature of ocean waves and try to point out their importance to the marine environment.

8.1 WAVES AS ENERGY

It is stated in the Appendix that radiant energy is propagated in the form of electromagnetic waves and that the electromagnetic spectrum (Fig. A-6) illustrates the tremendous range in wavelengths associated with this phenomenon. Throughout the world ocean, waves that move over the surface and within the sea are also a form of energy—mechanical energy. There is always some sort of wave motion on the surface of the ocean. This motion represents propagation of mechanical energy along a density discontinuity (the sea surface). The energy is imparted by wind, earthquakes, volcanos, landslides, meteorological phenomenon, or even other

118

waves. Waves extend not only from the air-sea interface but also along any pycnocline. As in the electromagnetic spectrum, there is a complete spectrum of sea waves, ranging from small capillary waves having a wavelength of about 2 centimeters to the tides with wavelengths of thousands of kilometers.

8.2 GENERAL FEATURES AND DESCRIPTION OF WAVES

In order to describe and discuss waves, we must first introduce descriptive terminology. A schematic profile of an ideal surface wave and its component parts is presented in Fig. 8-1.

Most waves move in what is called the *direction of propagation.* The *celerity* is the speed with which a crest or trough moves in the direction of propagation. The *period* of a wave is the interval of time required for successive crests or troughs to pass a fixed point; for example, successive crests of waves with a 8-second period will pass a fixed point every eight seconds.

The length, period, and celerity of a wave are related. For all progressive (i.e., moving) waves, this relationship can be demonstrated in the formula:

$$C = \frac{L}{T} \qquad\qquad [8\text{-}1]$$

where C = celerity or speed, L = wavelength, and T = period. Note that this formula does not state which factors govern the velocity of a progressive wave. It merely relates speed and the wave characteristics.

8.3 CLASSIFICATION OF WAVES

Sea waves are generally classified in three ways: (1) by the apparent shape of the sea surface, (2) by the relationship of the waves to the depth of water, and (3) by wave origin. In this text, the subject of ocean waves

FIGURE 8-1

Component parts of an ideal water wave consist of speed (C), wave length (L), height (H), crest and trough.

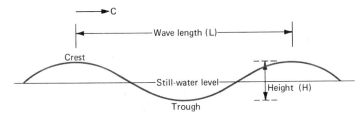

is discussed mainly in terms of origin; however, it is sometimes necessary to use the other two classifications.

The first classification deals with the shape of the wave form (or sea surface). If the wave appears to propagate in a definite direction, it is said to be a *progressive wave* (Fig. 8-1). An example of a progressive wave train is the motion that results when a rock is dropped in a quiet pond. If a wave appears to be a vertical undulation of the sea surface and does not propagate, it is called a *standing wave* (Fig. 8-2). Most waves have properties of one of these two classes.

The second classification deals with the relationship of the wave form to the depth of the water. If the water is deeper than ½ the wavelength of a given wave, that particular wave is said to be a *deep-water wave*. On the other hand, if the depth is less than about ½₀ the wavelength, the wave is considered a *shallow-water wave*. This distinction is made because a wave in shallow water is affected by the bottom and therefore has different characteristics than the same wave in deeper water. In the transition zone, where the depth is between ½ and ½₀ of the wavelength, the wave characteristics change from deep-water to shallow-water types. This relationship is illustrated by Fig. 8-3.

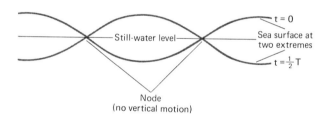

FIGURE 8-2

Idealized representation of a standing wave. The wave form appears to undulate vertically rather than move in a particular direction.

FIGURE 8-3

In shallow water the characteristics of a wave change with respect to the water depth.

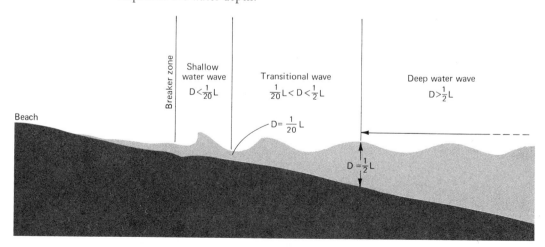

FIGURE 8-4
Small capillary waves having a wavelength on the order of a centimeter are
the first waves that form under the influence of the wind.

8.4   WIND WAVES

The waves familiar to most people are those caused by the wind.
Waves of this origin occur in many sizes and shapes, ranging from a few
centimeters in height and length to hundreds of meters in length and
over 30 m high.   Where these waves are present, the sea surface can vary
from a chaotic surface of breaking waves to a smooth plane interrupted
only by a systematic undulation.   Let us illustrate the complete spectrum
of what can be called *wind waves*.   To study their origin, propagation, and
breaking characteristics, we shall follow a group of waves from the time
they are generated within a storm at sea to the time they break against a
beach thousands of kilometers away.

GENESIS AND METAMORPHOSIS OF WIND WAVES
As the wind begins to blow over the smooth surface of the ocean, a
certain amount of energy of the wind is imparted by friction and pressure
fluctuations to the sea surface.   This energy is partly manifested in the
form of waves.   The harder the wind blows, the greater the transfer of
energy and the greater the waves.   The size of waves (i.e., their height
and wavelength) is determined not only by the force of the wind, but
also by its duration and by the distance over which it blows, or its *fetch*.
For example, a 15.3 m/sec (30-knot) wind can produce a wave about 4
meters high.   However, a 4-meter wave does not originate the instant that
the wind begins to blow.   The first waves to form are *capillary waves* (Fig.

121

TABLE 8-1

MINIMUM FETCH AND DURATION REQUIRED FOR FULL
DEVELOPMENT OF WAVES ASSOCIATED WITH VARIOUS
WIND SPEEDS (Data from Wolfe et al. Earth and
Space Sciences, 1966, D.C. Heath and Co.)

Wind	Fetch	Duration
m/sec (kn)	(km)	hr
5.1 (10)	18.5	2.4
10.2 (20)	140	10
15.3 (30)	520	23
20.4 (40)	1320	42
25.5 (50)	2570	69

8-4). These waves, in turn, make the sea somewhat rougher and allow more efficient interaction between the wind and the sea surface. There is then an increase in the transfer of energy from the wind to the sea. After the 15.3 m/sec wind has blown for about 5 hours, the larger waves that have formed are approximately 3 meters high; after 23 hours, the waves have reached an average height of 4.3 m. No matter how long the wind continues to blow beyond 23 hours, the waves will not become higher, because the dissipation of energy by viscosity is equal to the energy imparted to the sea by the wind. For the given conditions, therefore, some sort of dynamic equilibrium is attained.

Since these waves are progressive and are in motion, the distance traveled by the growing waves before an equilibrium condition is reached is about 180 kilometers for the 15.3 m/sec wind. The speed of these waves approaches the speed of the wind.

It can be said that duration and fetch are limiting; that is, for a given wind, a certain duration and fetch are required in order for fully developed waves to occur. Beyond this required time and distance, the waves grow no higher. If, however, either variable is limited, the maximum wave form will not be attained. Table 8-1 shows the minimum fetch and duration required for a fully developed wave at varying wind speeds. The characteristics of fully developed wind waves are given in Table 8-2.

In a storm area where the sea surface is turbulent and filled with spray, whitecaps, and breaking waves, a whole spectrum of waves is present. There are not only the "mature" waves resulting from the particular wind conditions, but also many sizes of waves traveling in different directions (Fig. 8-5). Because storm centers generally move at slower speeds than the larger wind waves they produce, the waves propagate from this

TABLE 8-2

CHARACTERISTICS OF FULLY DEVELOPED WIND WAVES
(Data after Wolfe et al. Earth and Space Sciences, 1966, D.C. Heath & Co.)

Wind speed	Average period	Average length	Average height	Maximum height*	Approx. celerity
m/sec (kn)	sec	m	m	m	m/sec (kn)
5.1 (10)	2.9	8.5	0.27	0.55	4.6 (9)
10.2 (20)	5.7	32.9	1.5	3.0	8.7 (17)
15.3 (30)	8.6	76.5	4.1	8.5	13.3 (26)
20.4 (40)	11.4	136.0	8.5	17.3	17.8 (35)
25.5 (50)	14.3	212.0	14.8	30.0	21.9 (43)

* Average of the highest 10 per cent of all waves.

area in all directions, often traveling for thousands of kilometers before being checked by friction or obstructed by a shoreline.

As the waves propagate away from the storm area, the sea surface begins to change. Spray, whitecaps, and breaking waves soon disappear; to an observer, the sea surface has a smoother appearance than within the storm area. Waves of different length disperse in such a way that the longer, therefore faster, waves move ahead of the slower, shorter waves. At progressively greater distances from the storm edge, the total wave train

FIGURE 8-5
Under the influence of strong winds the sea surface takes on a chaotic shape composed of many sizes of waves travelling in many directions. (Photograph courtesy of Peter B. Taylor).

continues to disperse until the longer waves may be traveling several days ahead of the shorter waves. Here the sea surface appears smoothly interrupted by waves having exceedingly uniform height and wavelength. These waves are of the progressive type and are referred to as *swell* (Fig. 8-6). Of course, all gradations exist between the chaotic sea state existing in a storm, and swell conditions occurring thousands of miles from the generating area.

Two types of motion are associated with swell: that of the wave form and that of the water. It is important to realize that waves transport only energy, not water. Although the form of the progressive wave (i.e., the actual hump in the sea surface) moves across great expanses of the ocean at relatively high speeds, the water does not accompany the wave form in its travel. Actually, the individual water particles travel in vertical circular orbits (Fig. 8-7a). With each complete revolution, energy is transferred to the water lying adjacent in the direction of wave propagation. Thus, the water itself only revolves. It is the energy associated with a given wave that is passed on continuously from water particle to water particle and in this way propagates as the wave forms.

To verify this fact, note how a Ping-Pong ball behaves on the surface of the sea. As a wave passes, the ball moves in a verticle circle as it

FIGURE 8-6
Aerial view looking down on a large ocean swell on the sea surface. Approximate wavelength is 75 meters. (Photograph courtesy of Barbee Scheibner).

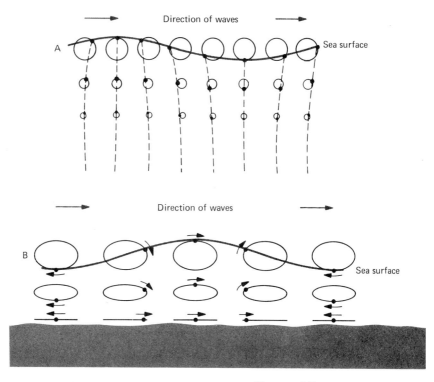

FIGURE 8-7
The motion of water particles as a wave passes are circular in deep water
(A); and flattened ellipses in shallow water (B). The depth of wave
disturbance is approximately ½L.

receives and transmits the energy of the waves (Fig. 8-7A). The motion
of a similar ball floating below the surface is the same. The only difference between a ball at the surface and one at some depth is that the
radius of the circular orbit decreases as the depth increases. At a depth of
approximately one-half the wavelength, the orbital radius is approximately
4 per cent of its surface value—a negligible figure. It is said, therefore,
that a wave does not exist below a depth of about one-half its wavelength.
In other words, a deep-water wave "feels bottom" at a depth of half its
wavelength.

The speed (C) at which a deep-water wave progresses can be found
by the following equation:

$$C^2 = \frac{gL}{2\pi} \tag{8-2}$$

where C = celerity, g = acceleration due to gravity, L = wavelength, and

$\pi = 3.14$. If we combine this equation with Eq. (8-1), we obtain:

$$C = \frac{gT}{2\pi} \qquad\qquad [8\text{-}3]$$

or

$$L = \frac{gT^2}{2\pi} \qquad\qquad [8\text{-}4]$$

Therefore, a deep-water wave having a period of 10 sec has a velocity of:

$$C = \frac{gT}{2\pi} = \frac{(980 \text{ cm/sec}^2)\,(10 \text{ sec})}{(2)\,(3.14)}$$

$$= 1{,}560 \text{ cm/sec} = 15.6 \text{ m/sec} \qquad [8\text{-}5]$$

and a wavelength of:

$$L = CT = 15.6(10) = 156 \text{ m} \qquad\qquad [8\text{-}6]$$

Note that the speed of swell depends upon the square root of its wavelength (Eq. 8-2). Consequently, in deep water, longer waves move faster or overtake short waves as they propagate from the storm area. We call this process of wave separation *dispersion*. Because of dispersion, the first waves to reach a distant beach from a large storm are the longer waves.

Where the depth of water is shallower than one-half the wavelength, a deep-water wave feels bottom and starts the transition to a shallow-water wave. As this happens, the circular orbits of water particles become flattened into ellipses (Fig. 8-7B) and the speed of waves begins to decrease. Since the wave period remains constant, the length of these waves must also decrease. As the length decreases, the steepness (i.e., the ratio of height to length) increases. When the steepness increases so that the ratio of height to length exceeds 1 to 7, the wave becomes unstable and breaks, forming surf, thus finally expending its energy on some distant beach (Fig. 8-8).*

The speed of a shallow-water wave is found by the following formula:

$$C = \sqrt{gD} \qquad\qquad [8\text{-}7]$$

---

* Chapter 10, Inshore Oceanography, further discusses *shoaling waves*, which influence nearshore sedimentary processes.

where D = depth. No dispersion is associated with shallow-water waves, because their speed depends upon depth rather than length or period.

8.5 CATASTROPHIC WAVES

TSUNAMI

A *tsunami*, or seismic sea wave (often referred to as a "tidal wave," although it has nothing to do with the tides) is a shallow-water sea wave that is caused by submarine crustal movements or the sudden displacement of a large volume of seawater. These waves originate from any of several geologic phenomena: submarine volcanic explosions, submarine landslides, earth movements (faulting), or even large landslides into the sea.

The physical characteristics of a tsunami are striking. It is because their wavelengths are between 120 and 720 km that they are shallow-water waves. Their speed of propagation, therefore, is determined by the water depth (see Eq. 8-7). In the open ocean (where the average depth is 4,000 m), they can propagate at about 200 meters per second or 400 nautical miles per hour. Consequently, a tsunami caused by submarine faulting

FIGURE 8-8

Waves travelling great distances ultimately expend their energy on the shorelines of the world. Where the beach is shallow the surf zone extends far offshore. On steep beaches the surf zone is very narrow.

in the Aleutian Islands would require only 5 hours to travel to Hawaii. In the open ocean, however, a tsunami could pass without being observed, because its height is only about 30 centimeters and its period varies from 10 to 60 min. But as it approaches shore, such a wave can attain heights of 35 m.

Because of the origin and nature of tsunamis, they have been known to cause great damage and loss of life to certain coastal areas throughout the world. Probably the most widely publicized example of the magnitude and destructive force associated with these waves is that caused by the volcanic explosion of the island of Krakatoa in the Sunda Strait on August 26 and 27, 1883. The resulting waves reached heights as great as 35 m in some of the Indonesian islands. Approximately 32 hours after their formation, the waves were detected in the English Channel; however, their height had decreased to several centimeters, since they had traveled almost 20,000 kilometers.

The rim of the Pacific Ocean is a seismically active area, so the occurrence of earthquakes and tsunamis has been relatively frequent there. One of the factors contributing to the great loss of life caused by these seismic sea waves is the fact that their occurrence cannot be predicted. They arrive on distant shores without much warning. Sometimes a trough reaches shore first, causing an outward rush of water that lowers the water surface to an abnormal degree. Several minutes later, the first of a series of large crests and deep troughs arrives. Each is separated by several minutes, and the entire train of waves lasts several hours. Since 1946, a tsunami-warning system consisting of a communications network was set up to transmit warnings of an impending tsunami to Pacific area countries and islands. This system is not the final answer, however, for tsunamis do not occur after all earth tremors or movements. The presence of an earthquake does not necessarily signify a tsunami. Moreover, the exact cause of the tsunami is not known. They seem to be related to both vertical and horizontal displacements of the sea floor as well as submarine slumping, but there may be other causes. Possibly it will be necessary to place electronic sensing devices on the sea floor to signal automatically the passage of a tsunami—no matter what its cause or origin.

STORM TIDES

Storm tides are more or less solitary occurrences of abnormally high water levels caused by a combination of meteorological and oceanic conditions. The primary cause of these waves are violent storms that sweep great amounts of ocean water against coastal regions. The occurrence of a *storm setup* in itself may not have a great damaging effect on a coastal region. Coupled, however, with other phenomena such as strong winds, large wind waves, high tides, and uncommonly low atmospheric pressure (causing slight sea-level increases), storm tides can inundate coastlines and cause great devastation and loss of life.

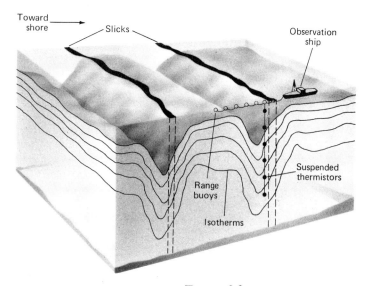

FIGURE 8-9

Block diagram showing an internal wave propagating toward shore off San Diego. (After Neumann & Pierson, 1966; by permission of Prentice-Hall, Inc.).

For example, storm tides as high as 5 m have struck the Texas coast. In 1900, the city of Galveston, Texas, was nearly destroyed. A combination of hurricane-force winds, storm setup, high waves, and low pressure caused flooding of the city, with property damage running into tens of millions of dollars and over 5,000 people killed.

8.6   INTERNAL WAVES

Internal waves are similar to ordinary sea waves except that they occur within the sea rather than at the surface. They exist at density discontinuities between water layers, especially at the pycnocline. In a homogeneous sea, they cannot exist. Internal waves are so long they must be detected by the periodic rise and fall of isotherms, which are lines connecting equal temperatures on oceanographic data profiles. Figure 8-9 shows how several isotherms vary with depth and time. In this particular example, the period of the oscillations is about 15 minutes and the amplitude is about 7 meters. *Slicks* that are seen on the surface of the ocean are often manifestations of internal wave movement and are generally located above a wave midway between trough and crest (Fig. 8-10).

Internal waves characteristically have greater amplitude and slower speeds of propagation than surface waves. Internal waves can also be found in both shallow and deep water. Table 8-3 illustrates the wide variation in the properties of these waves.

TABLE 8-3

GENERAL CHARACTERISTICS OF INTERNAL WAVES

	Period	Amplitude	Speed
Shallow	4 min–25 hr	2 m	5 cm/sec
Deep	4 min–25 hr	100 m	100 cm/sec

The causes of internal waves are varied and not completely understood. Early reports from steamship captains described conditions of "dead water" through which their vessels appeared to move sluggishly. In 1904, V.W. Ekman explained this phenomenon by postulating the existence of near-surface internal waves. A vessel that was traveling at about the speed of an internal wave (100 cm per sec) would have a significant part of its thrust utilized in generating internal waves at the lower bound-

FIGURE 8-10
Photo of surface features associated with internal waves. (Photograph courtesy of William McLeish).

ary of the relatively fresh upper layer of water—provided the depth of the boundary were about the same as the ship's draft. If the vessel increased its speed, the thrust would again be used to overcome the fluid resistance, and the ship would no longer appear to "stick" in the water.

By now, scientists have discovered that internal waves are, in fact, caused by many phenomena: storms, surface waves, ships, tidal action, wind blowing over the sea surface, or possibly even any transient disturbance that combines with these other conditions.

8.7   STANDING WAVES

In contrast to a progressive wave, where the wave form propagates in a given direction, a standing wave (also called a *seiche*) is characterized by a surface oscillation that does not appear to move horizontally. Standing waves occur in basins that are closed or almost closed to the ocean. Actually, the ocean basin itself can maintain a standing wave (see the discussion of tides in Chapter 9). The characteristics of a standing wave are related to the nature of the enclosing basin. When an impulse of energy is supplied to the basin of water, the water surface will oscillate up and down by tilting first in one direction and then in the opposite direction, as illustrated in Fig. 8-2. There is a place, called a *node*, where no vertical movement—only horizontal movement—occurs. Maximum vertical movement occurs at points called *antinodes*. Depending upon how, when, and where the energy impulse occurs, a basin may oscillate with several nodes (also illustrated in Fig. 8-2). The period of oscillation of a standing wave depends on the geometry (i.e., the length and depth) of a basin. This relationship is given for a single node in the following formula:

$$T = \frac{2L}{\sqrt{gD}} \tag{8-8}$$

where T = natural period of oscillation, L = length, g = acceleration of gravity, and D = depth.

If energy impulses are supplied to a basin periodically so that the period of the impulses is about equal to the natural period of the basin, the height of the standing wave will increase with time. In other words, the basin will resonate, and the vertical motion of the water at the antinodes will increase. In nature, energy impulses are supplied to the basins by storm surges, sudden changes in barometric pressure, progressive waves, earthquakes, tides, and such transient impulses as landslides or even thermonuclear explosions. The sudden single impulse causes a series of oscillations that is gradually damped by boundary friction and viscosity. Such oscillations can be catastrophic if the impulse is large. More severe, however, are cases produced by periodic impulses, like waves and tides, if they can produce enough resonance to cause the oscillations to build up to dangerously high changes in the sea level at the antinodes (i.e., the shores of the basin).

There is also a kind of single-oscillation wave called a *landslide surge*. A landslide falling into a basin may displace large volumes of water, causing an oscillation that can have devastating effects upon the basin and surrounding areas. Such a landslide surge occurred in Lituya Bay in Alaska in 1958 where 40 million cu m of earth fell into the bay from elevations of 1,000 m. Water rushed 500 m up the mountains on the opposite side of the bay, and a water wave 15 m high surged seaward at a speed of 50 m per sec, carrying several vessels anchored within the bay over a bar located at the entrance. A similar disaster occurred in the Vaiont reservoir in Italy in 1963. A landslide containing 600 million tons of earth created a wave as high as 250 m that overtopped the dam. It swept into the village of Longarone and caused many casualties and great damage.

READING LIST

BASCOM, W., "Ocean Waves," *Scientific American*, CCI, No. 2 (August 1959), 74–78.

————, *Waves and Beaches: The Dynamics of the Ocean Surface*. Garden City, N.Y.: Doubleday & Co.. Inc., 1964. 267p.

BERNSTEIN, J., "Tsunamis," *Scientific American*, CXCI, No. 2 (August 1954), 60–64.

BIGELOW, H.B., AND W.T. EDMONDSON, *Wind Waves at Sea, Breakers and Surf*, U.S. Navy Hydrographic Office Publication 602. Washington, D.C.: Government Printing Office, 1947. 177p.

CHRISTOPHER, P., H. RUSSEL, AND D.H. MACMILLAN, *Waves and Tides*. London: Hutchinson's Scientific & Technical Publications, 1952. 348p.

ONE
TWO
THREE
FOUR
FIVE
SIX
SEVEN
EIGHT
NINE
TEN
ELEVEN
TWELVE
THIRTEEN
FOURTEEN
FIFTEEN

tides

For many of their activities, men must understand the origin and nature of tides and be able to predict the tidal fluctuations of the world ocean. Certain kinds of commerce and industry, reclamation of land from the sea, amphibious military operations, beach recreation, and solid waste disposal in coastal areas are all dependent upon the tides. The rise and fall of the tides is also important to ocean processes not directly related to man, particularly coastal morphology, chemical processes, and the life cycle of organisms in coastal regions.

9.1   THE GENERAL NATURE OF TIDES

A tide is the alternate rising and falling of the sea level as a result of the gravitational forces exerted on the earth by the sun and the moon. The regularity of this rise and fall, or *flood* and *ebb*, is associated with the periodicity of solar and lunar gravitational effects. The height of a tide is related to the sum of water displacements produced by these gravitational forces.

The tides propagating through the world ocean are strongly modified as they approach the shallow margins of the continents. However,

133

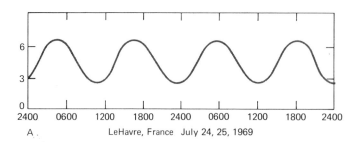

A. LeHavre, France July 24, 25, 1969

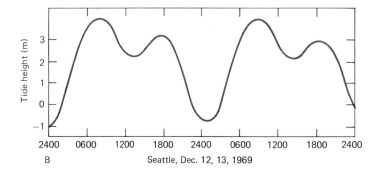

B Seattle, Dec. 12, 13, 1969

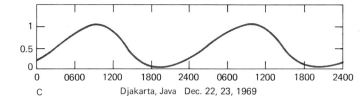

C Djakarta, Java Dec. 22, 23, 1969

FIGURE 9-1
These tide curves illustrate semi-daily (A), mixed (B), and daily (C) types.

they are modified with respect only to amplitude and velocity, not to frequency. The influence that a coast exerts on tidal characteristics is determined by coastal size, shape and bathymetry; therefore, no two locations in the world exhibit the same tidal behavior.

The most obvious tidal fluctuations are those that occur once daily (diurnal) or twice daily (semidiurnal). These fluctuations have either equal or unequal amplitudes, so a wide variety of tide curves is observed throughout the world. Some examples are shown in Fig. 9-1.

Tides with periods longer than 24 hours also occur. If we observe a tide curve over a lunar month, periodic variations in the *tidal range* (difference between successive high and low tides) become evident. Times of maximum tidal range, called *spring tides,* occur at about two-week intervals. Times of minimal tidal range, called *neap tides,* also have a two-week interval but lag behind spring tides by about one week. Some examples of spring and neap tides from various geographic locations are shown in Fig. 9-2. Notice that the tide at Manila is diurnal except during neap tides, which are semidiurnal (Fig 9-2C).

134

Tides flow as shallow-water progressive waves in the open water, but in coastal regions, they may act differently. In Chesapeake Bay and Puget Sound, for instance, the tide wave tends to progress as a shallow-water wave whose speed is a function of water depth (according to Eq. 8-7). In other regions, the tides act as standing waves. The most famous example is in the Bay of Fundy in Nova Scotia. Here, large vertical fluctuations occur at the head of a bay while at the mouth of the bay, the water level does not change appreciably. The maximum tidal range at the head of the Bay of Fundy is in excess of 15.4 m, whereas the tidal range at the mouth is approximately 3½ m.

Not only the ocean but also everything on the earth responds to the gravitational attraction of the sun and the moon. Tides occur in the atmosphere and in lakes. Lake Superior, for example, has a tide of a few centimeters. The earth's crust also responds to tidal forces. The nature of the tides, their magnitude, and periodicity are governed by the positions and motions of the bodies in our solar system. Before proceeding to consider tidal effects on the earth, we must become familiar with some of these motions.

FIGURE 9-2

Tides at Balboa, San Francisco, Manila and Pakhoi. The time in days is given on the bottom of the figure. The curves vary from semidiurnal (A); to mixed (B and C during neap tides); to diurnal (D).

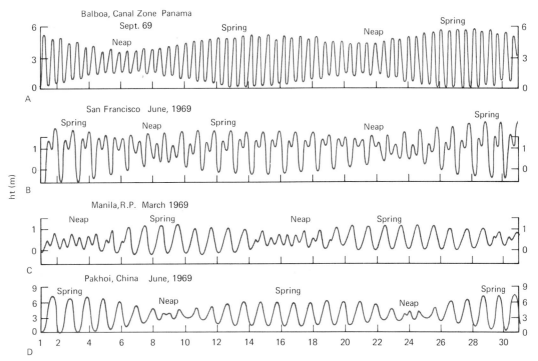

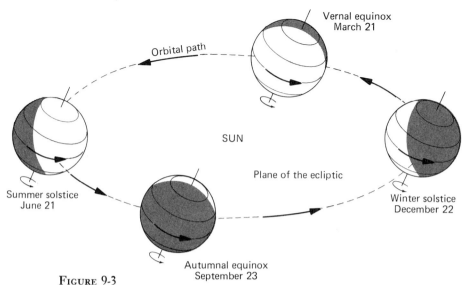

Orbital path

Vernal equinox
March 21

SUN

Plane of the ecliptic

Summer solstice
June 21

Winter solstice
December 22

Autumnal equinox
September 23

FIGURE 9-3

The four seasons as related to the earth's orbit around the sun.
The earth's axis remains fixed in space and is tilted approximately
23.5° with respect to the vertical (after A. N. Strahler, 1963.).

9.2 THE POSITION OF THE EARTH IN THE SOLAR SYSTEM

Almost everyone knows a few elementary facts about the earth and its association with the sun: The earth is approximately 149,642,000 km from the sun (93,000,000 miles); the earth completes one revolution about the sun in approximately one year; and the earth rotates around its axis in the time defined as one day. These three facts provide a basis for a further description of the earth-sun association that is less well known but nonetheless important. Let us study Fig. 9-3.

This figure illustrates schematically four positions of the earth during its yearly orbit around the sun (as viewed from above the north polar axis of the earth). Note the following facts:

1. The earth's orbit traces an ellipse with the sun at one of the foci. The distance between the two bodies is not constant.
2. The earth not only revolves around the sun in a counterclockwise direction; it also rotates around its own axis in a similar direction.
3. The axis of the earth is not oriented vertically with respect to the orbital plane; it is tilted about 23½ degrees from the vertical.
4. The axis of the earth always points in essentially the same direction relative to the stars. During part of the year, the northern hemisphere faces toward the sun; at other times, it faces away.

Because the earth's axis changes position relative to the direct rays of the sun, we have seasons and corresponding differences in the length of day and night. At the summer solstice in the northern hemisphere (about

136

June 21), the earth's axis has the greatest inclination toward the sun, and it is the northern hemisphere that receives the sun's rays most directly. This is the longest day of the year in this hemisphere. Indeed, north of the Arctic Circle, there are 24 hours of daylight. When it is summer in the northern hemisphere, however, it is winter in the southern. The shortest day of the year occurs here in June, and there are 24 hours of darkness south of the Antarctic Circle. At the summer solstice in the northern hemisphere, the noon sun is directly over the Tropic of Cancer (23½ degrees N latitude).

At the winter solstice in the northern hemisphere (about December 22), the conditions prevailing on the earth are, of course, the reverse of those at the summer solstice. It is the northern winter (with its corresponding short days) and the southern summer (with long days). At this time, the sun is directly over the Tropic of Capricorn (23½ degrees S latitude).

The vernal and autumnal equinox are the times of the year when day and night are of equal length over the whole earth. Day and night are always of equal length on the equator. At the vernal equinox (about March 21), it is spring in the northern hemisphere and autumn in the southern hemisphere. At the time of the autumnal equinox (about September 23), it is autumn in the northern hemisphere and spring in the southern hemisphere. The noon sun is directly over the equator at both the vernal and autumnal equinox.

9.3   THE TIDE-PRODUCING FORCE

The variety of tidal phenomena affecting the earth arises from forces associated with the motion of mutually attracted celestial bodies. The earth, the moon, and the sun are drawn together by forces of gravitational attraction. They are kept apart by the centrifugal force arising from their orbital motion. The relative position that these bodies maintain in space is determined by a balance of the opposing forces. The orbit of the moon about the earth and of the earth around the sun are paths along which the balance of forces is realized.

To discuss the tide-producing force systematically, we must consider the earth-moon system as different from the earth-sun system, and then consider how the force of gravitational attraction and centrifugal force are regulated. The concepts developed for one system will apply to the other system as well. We use the earth-moon system because its tide-producing force is over twice as great as that of the sun. Once the effect of the earth-moon system is understood, the influence of the sun can be added.

The earth and the moon revolve around their common center of gravity, called the *barycenter*, in a counterclockwise direction when viewed from the north polar Axis. The moon completes one revolution around the earth every 29.53 days. This motion represents a dynamic equilibrium; that is, the gravitational attraction tending to draw the earth and the moon together is balanced by the centrifugal force arising from their orbital motion about one another. Figure 9-4 shows how the centrifugal and gravitational force vectors are seen to balance at the center of the earth.

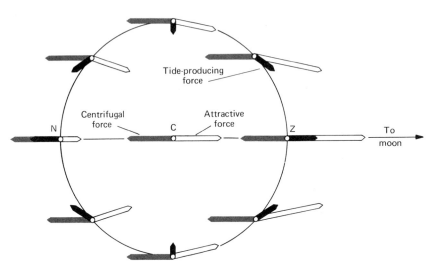

FIGURE 9-4
Distribution of centrifugal and gravitational (attractive) forces acting on the Earth. (Modified after A. N. Strahler.)

Let us now investigate the distribution of gravitational and centrifugal forces on the surface of the earth to find the reasons for tidal behavior in the ocean.

THE FORCE OF GRAVITATIONAL ATTRACTION

One of Newton's laws of motion relates the force of gravitational attraction between two bodies to the product of their mass divided by the square of the distance between them:

$$F_g = \frac{GM_1M_2}{r^2} \qquad\qquad [9\text{-}1]$$

where M_1 and M_2 = masses of attracting bodies, $r =$ distance between objects, and $G =$ gravitational constant.

For the earth-moon system, the product M_1M_2 is constant so Eq. (9-1) can be expressed as:

$$F_g = \frac{M}{r^2} \qquad\qquad [9\text{-}2]$$

where $M = GM_1M_2 =$ a constant. Equation (9-2) describes how the earth is influenced by gravitational forces acting between the earth and the moon. The forces are distributed as shown in Fig. 9-4. On the side of the earth nearest the moon, the distance (r) between point Z and the center of the moon is less than from the center of the earth (point C), so F_g is greater at Z than at C. Point N is farther from the moon than point C; therefore, F_g at point N is smaller than at the center. These differences are small but important, because they lead to the unbalanced forces that are responsible for the tides.

CENTRIFUGAL FORCE

Centrifugal force is the other major force producing tides. Centrifugal force results from the motion of a mass along a curved path. In the case at hand, we are interested in the centrifugal force on the earth as it travels in its orbit about the center of the earth-moon system. The force is directed away from the center of the orbit and may be expressed as:

$$F_c = \frac{MV^2}{r} \qquad\qquad [9\text{-}3]$$

where $M =$ the mass of the orbiting body, $r =$ the radius of curvature of its orbital path, and $V =$ the velocity of motion. The value of F_c is the same at all points on the earth, because the orbit of any point on the earth has the same radius as the orbit of any other point of curvature.

We can clarify this fact by considering the analogy of a Ferris wheel (Fig. 9-5). The axis of the Ferris wheel represents the center of the earth-moon system and the gondola is the earth. The orientation of the gondola at different positions is indicated by the seats. If there is no frictional coupling between the Ferris wheel and the gondola, the gondola will remain oriented with respect to the ground as the Ferris wheel revolves. Note that no matter what position is chosen, the orbits of all points have the same radius ($r_1 = r_2 = r_g$). The centrifugal force acting on all points on the gondola is the same, since this force, F_c in Eq. (9-3), depends only upon the radius if the mass and velocity are constant. The distribution of F_c on the earth is constant and unidirectional, because the motion of the earth about the barycenter is similar to that of the Ferris wheel.

It is not necessary to include in this discussion the centrifugal force arising from the rotation of the earth about its own axis. The earth's shape is in dynamic equilibrium with respect to its internal gravitational and centrifugal forces; consequently, there is no resultant force to influence tidal phenomena.

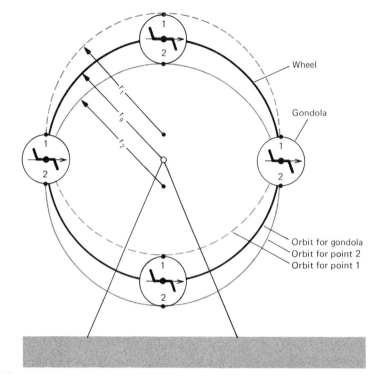

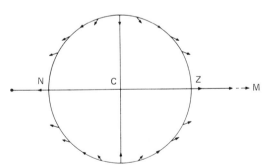

FIGURE 9-5
Schematic representation of a Ferris wheel in which there is no frictional coupling between the gondola and the wheel. The gondola maintains its position with respect to the ground at all times. The orbits traced by various points on the gondola as it revolves are also included.

FIGURE 9-6
Distribution of tide-producing force on a plane of the earth oriented toward the moon (after Sverdrup et al., 1942).

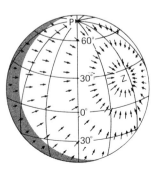

FIGURE 9-7
Distribution of the tractive force over the surface of the earth. The point Z is facing the moon (After Sverdrup et al., 1942; by permission of Prentice-Hall, Inc.).

The *tide-producing force* is obtained by resolving the combined effects of the centrifugal and gravitational forces throughout the earth. This resolution is shown in Fig. 9-6. Note that two regions of intense tide-producing forces exist: one facing the sublunar point (where gravitational forces dominate) and one on the opposite side of the earth (where centrifugal forces dominate). The surface resultant of the tide-producing force is called the *tractive force* (Fig. 9-7). Tractive forces cause ocean water to accumulate, producing *two* tidal highs, one at the sublunar point and another on the opposite side of the earth.

The moon circles the earth once every 29.53 days; during the 24 hours of one rotation of the earth on its axis, the moon completes $\frac{1}{29.53}$ of its revolution about the earth. As a result, the moon appears to return to a point directly overhead in a period slightly longer than one day. The interval is 24 hrs and 50 min, the so-called *lunar day*. This phenomenon is illustrated schematically in Fig. 9-8. Note that two high and two low tides sweep the circumference of the earth within the period of the lunar day. Such a tide is called a *lunar semidiurnal tide* and is illustrated in Fig. 9-1A.

Further complexities arise when we try to describe the motion of the tidal bulges with respect to the earth's surface. As the moon revolves around the earth, its declination changes constantly throughout the lunar month. The plane of the moon's orbit is tilted 5 degrees, 9 minutes, with respect to the plane of the ecliptic (Fig. 9-9). Hence, during each lunar month, the declination of the moon's orbit changes continually between positions 28½ degrees above and below the equator. As declination varies, different places on the earth exhibit different tidal behavior. During the time of the month when the moon is at maximum declination, every point on the earth exhibits a lunar semidiurnal tide with a period of 12 hr

FIGURE 9-8
Cause of lunar semi-diurnal tide constituent.

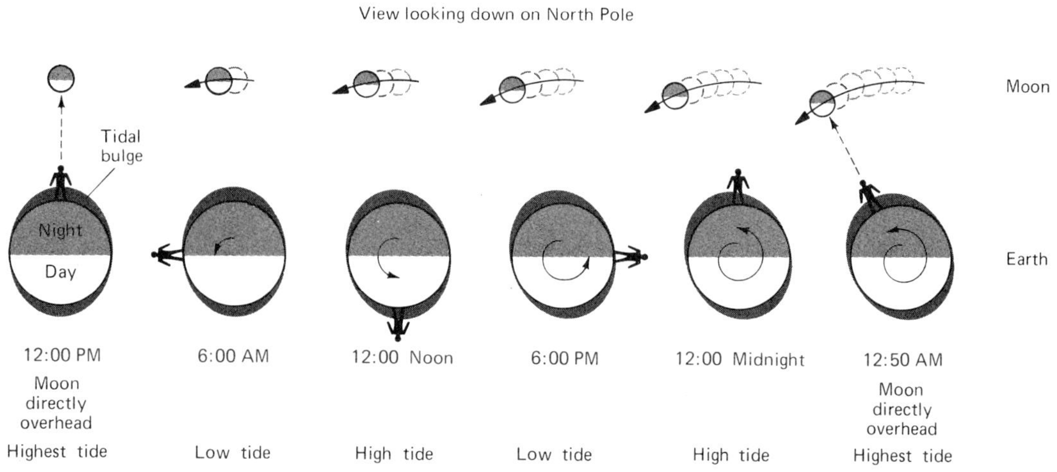

View looking down on North Pole

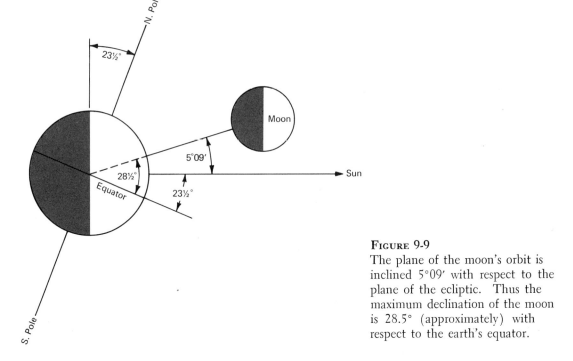

Figure 9-9
The plane of the moon's orbit is inclined 5°09′ with respect to the plane of the ecliptic. Thus the maximum declination of the moon is 28.5° (approximately) with respect to the earth's equator.

and 25 min. Only at the equator do high and low tides have equal amplitude. Elsewhere, an inequality exists in extreme high and low tidal amplitudes. Tides of this type are called *tropic tides* (Fig. 9-10A).

The tidal curve in Fig. 9-10 A actually consists of a diurnal and semidiurnal component (Fig. 9-11). No purely diurnal tide exists, but the effect of declination is to add, or superimpose, a small diurnal component onto the existing semidiurnal pattern. The result is a mixed tide in which successive high and low water levels are different. Likewise, every unique movement of the sun or moon adds another tidal constituent onto the basic pattern—a phenomenon that will be discussed in the next section.

The lunar plane shifts to coincide with the earth's equator twice each lunar month. At that time, a somewhat different tidal configuration occurs (Fig. 9-10B). The tidal bulge is oriented symmetrically around the equator, and all portions of the globe influenced by tidal fluctuations are characterized by semidiurnal lunar tides of equal amplitude. Tides of this type are called *equatorial tides* (Fig. 9-10B).

This idealized and, in some ways, unrealistic portrayal of the tides emphasizes several of their important aspects. First, tide-producing forces are related to the combination of gravitational and centrifugal forces within the earth-moon system. Secondly, two tidal bulges are predicted by this analysis, which agrees with tidal observations. Finally, many tidal configurations are possible, depending on latitude and declination of the moon.

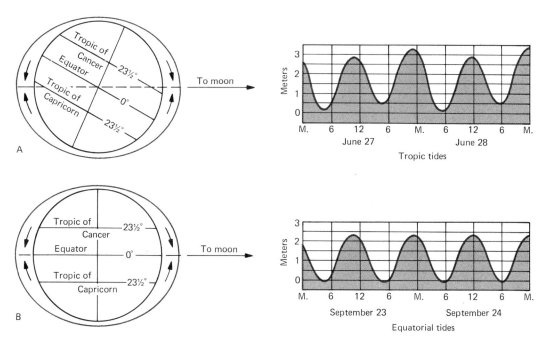

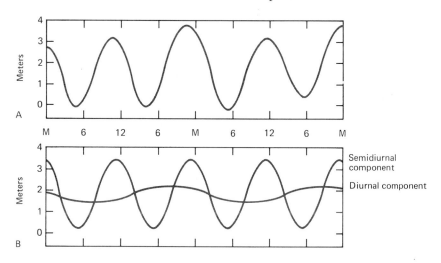

FIGURE 9-10

The tidal bulges shift with the declination of the moon causing Equatorial
and Tropic tides. The corresponding tidal curves are also shown.

(After A. N. Strahler, 1963).

FIGURE 9-11

Separation of a Tropic Tide curve (from Fig. 9-10A) into a
diurnal and semidiurnal component.

The identical analysis can be applied to water-level changes produced by the earth-sun system. Certain minor modifications, however, are required: (1) the period of the solar-diurnal tide is 24 hours and of the solar semidiurnal tide, 12 hours; (2) the plane of the sun's ecliptic is inclined away from the earth's equator. This inclination reaches a maximum of 23½ degrees on an annual cycle.

Actually, tidal bulges are shallow-water waves that cannot travel at the speed of the earth's rotation. The result is that the tides are *forced waves*: their period is dependent on the periodic motions of the moon and the sun, but their speed depends on the water depth (approximately 200 meters per second in the open ocean). As a consequence, high tide does

FIGURE 9-12
Relationship between the positions of the Moon and Sun with respect to the Earth (A) and the occurrence of Spring and Neap tides as shown in a plot of tidal range over a 30-day interval (B). (After A. N. Strahler, 1963).

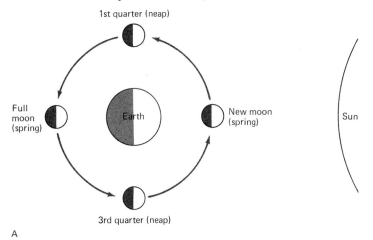

A

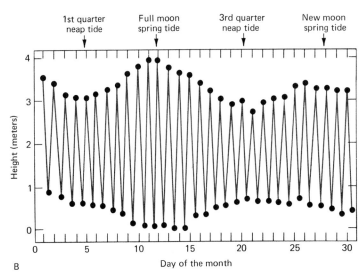

B   Day of the month

not necessarily coincide with the passage of the sun or the moon. The delay time for the moon is called the *lunitidal interval*. This interval remains approximately constant for a given locality but varies greatly throughout the world.

At certain positions in the moon's orbit around the earth, the tide-producing forces of the moon and the sun combine to cause a very large tidal range, called a *spring tide*. Spring tides occur when the earth, the moon, and the sun are in alignment (*syzygy*), as shown in Fig. 9-12A. Such a configuration occurs every half-revolution of the moon around the earth, or approximately every two weeks. When the moon is 90 degrees from the line connecting the earth and the sun (the *quadrature*), the lowest tidal range, the *neap tide*, occurs. These tides are also about two weeks apart, but they occur one week after each spring tide. Spring-neap variations are called the *lunar fortnightly constituents* of the tide-producing force. The relation to the phases of the moon can be seen in Fig. 9-12B. Spring tides occur at full and new moons, whereas neap tides occur at the first quarter and the third quarter of the moon.

9.4   TIDAL PREDICTION

When Sir Isaac Newton propounded his theories on gravity, he also introduced the *equilibrium theory of tides*. He assumed a water-covered earth of constant depth that remained in a state of equilibrium with regard to the interplanetary forces of gravity. Accordingly, the great tidal bulges should exactly follow the transit of the sun and the moon across the sky.

However, the tidal behavior envisioned in the equilibrium theory is inaccurate. Although the gravitational influences of the sun and the moon are evident on the earth, the speed at which the tidal bulges traverse the equatorial oceans is less than one-half the speed of the diurnal motions of the sun and the moon. Thus, the equilibrium theory does not deal correctly with the dynamics of water set in motion by the tractive forces.

Once the transport of water is considered, there arise several other complexities that further diminish the applicability of the equilibrium theory. In the first place, it is impossible to set the ocean waters in motion at a speed necessary to maintain equilibrium with the passage of the moon, because the quantity of water to be transported is too great. Secondly, once water is moving, it is influenced by local factors such as the shape, size, and depth of the ocean basin. These factors modify tidal motion. Tidal friction and meteorological influences are also important in influencing water motion in the world ocean.

The equilibrium theory is useful only for predicting tide-generating forces and some of the basic tidal configurations that are observed. In addition, spring and neap tides and other long-term tidal variations can be predicted, for they occur at a rate slow enough to agree with the theory. But still, actual local tidal motions on the earth cannot be calculated by using the equilibrium theory. As a result, mechanical techniques have

TABLE 9-1

THE MOST IMPORTANT CONSTITUENTS OF THE TIDE-GENERATING FORCE
(CONSTITUENT TIDES) (After Defant, 1958)

	Symbol	Period in solar hours	Amplitude $M_2 = 100$	Description
Semidiurnal tides	M_2	12.42	100.00	Main lunar (semidiurnal) constituent
	S_2	12.00	46.6	Main solar (semidiurnal) constituent
	N_2	12.66	19.1	Lunar constituent due to monthly variation in moon's distance
	K_2	11.97	12.7	Soli-lunar constituent due to changes in declination of sun and moon throughout their orbital cycle
Diurnal tides	K_1	23.93	58.4	Soli-lunar constituent
	O_1	25.82	41.5	Main lunar (diurnal) constituent
	P_1	24.07	19.3	Main solar (diurnal) constituent
Long-period tides	M_1	327.86	17.2	Moon's fortnightly constituent

been devised to predict the tides. These techniques require that tides be measured in any given place before prediction is possible.

The first step in tidal prediction is a systematic description of the periodic fluctuations of the tide-producing force. For this procedure, the complex orbital motions of the moon and the sun are separated into distinct cycles, or *tidal constituents*. Periodic fluctuations of the tractive forces reflect the combined periodicities of these motions.

Consider, for instance, the earth's revolution about the sun. If the variations in the earth's distance and declination with respect to the sun are neglected, then we could expect a purely semidiurnal tide. However, the distance from the earth to the sun varies; it is minimal during the northern winter (perihelion) and maximal during the northern summer (aphelion). These differences cause the tide-producing force to fluctuate on a yearly cycle. Hence, a perturbation, or tidal constituent, must be added to the basic pattern of the semidiurnal tide.

The declination of the sun also varies on an annual basis. During the summer solstice, the sun's declination reaches 23½ degrees N latitude; at the winter solstice, it reaches 23½ degrees S latitude. These declinational changes introduce another tidal constituent, a diurnal one that has a semiannual periodicity. The constituent for the effect of solar declination is analogous to the constituent showing the moon's declination (illustrated in Fig. 9-10).

This illustration, although highly simplified, shows how the tide-producing force can be related to the motion that the earth and the moon have with respect to the sun. Individual tidal constituents are sinusoidal and therefore are simple mathematically. Most importantly, when they are superimposed, a realistic representation of the tide-producing force is achieved. By this technique, called *harmonic analysis*, scientists can describe the combined effect of over 70 tidal constituents having periods from 12 hr to 1,600 years. Some examples are given in Table 9-1. Each tidal constituent is defined by a separate component of the motion of the earth and moon about the sun. Each component contributes a unique tidal fluctuation called a *partial tide*. The sum of partial tides gives the total tidal behavior. Although a great number of tidal constituents exist, not all have influence on every part of the ocean. Depending on its size, shape, and depth, each embayment, coastal area, and even each ocean basin responds to different constituents. For example, the Atlantic Ocean basin responds most strongly to forces that produce a semidiurnal tide, so lunar semidiurnal tides prevail there (Fig. 9-1A). The Pacific Ocean basin is large enough to respond to both diurnal and semidiurnal periods; hence, the tides along the Pacific shores reflect a mixture of diurnal and semidiurnal components and are called *mixed tides* (Fig. 9-1B). Tides that

FIGURE 9-13
Comparison between the computed and observed tides at Pula (January 6, 1909). The thin curves represent the 7 main constituent tides, the thick curve shows their resultant, and the thick broken curve the observed tide. (After Defant, 1958; by permission of the University of Michigan Press)

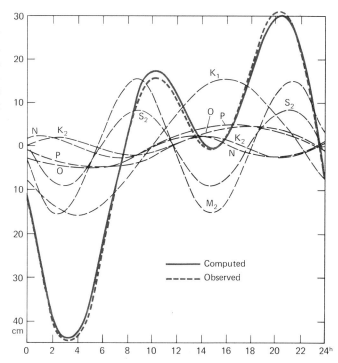

TABLE 9-2

SOME TIDAL CONSTANTS IN THE ATLANTIC (After Defant, 1958)

Place	Lat.	Long.	Amplitude in cm				Phase in degrees			
			M_2	S_2	K_1	O	M_2	S_2	K_1	O
St. John's (Newfoundland)	47°34' N	52°41' W	35.7	14.6	7.6	7.0	210	254	108	77
New York (Sandy Hook)	40°28'	74°01'	65.4	13.8	9.7	5.2	218	245	101	99
St. George (Bermuda)	32°22'	64°42'	35.5	8.2	6.4	5.2	231	257	124	128
Port of Spain (Trinidad)	10°39'	61°31'	25.2	8.0	8.3	6.7	119	139	187	178
Pernambuco	8°04' S	34°53'	76.3	27.8	3.1	5.1	125	148	64	142
Rio de Janeiro	22°54'	43°10'	32.6	17.2	6.4	11.1	87	97	148	87
Buenos Aires	34°36'	58°22'	30.5	5.2	9.2	15.4	168	248	14	202
Moltke Harbor (S. Georgia)	54°31'	36°0'	22.6	11.7	5.2	10.2	213	236	52	18
Capetown	33°54' S	15°25'	48.6	20.5	5.4	1.6	45	88	127	243
Freetown	8°30' N	13°14'	97.7	32.5	9.8	2.5	201	234	334	249
Puerto Lux (Las Palmas)	28°9'	25°25'	76.0	28.0	7.0	5.0	356	19	21	264
Ponta Delgada (Azores)	37°44'	25°40'	49.1	17.9	4.4	2.5	12	32	41	292
Lisbon	33°42'	9°8'	118.3	40.9	7.4	6.5	60	88	51	310
Brest	48°23'	4°29'	296.1	75.3	6.3	6.8	99	139	69	324
Londonderry	55°0' N	7°19' W	78.6	30.1	8.2	7.8	218	244	181	38

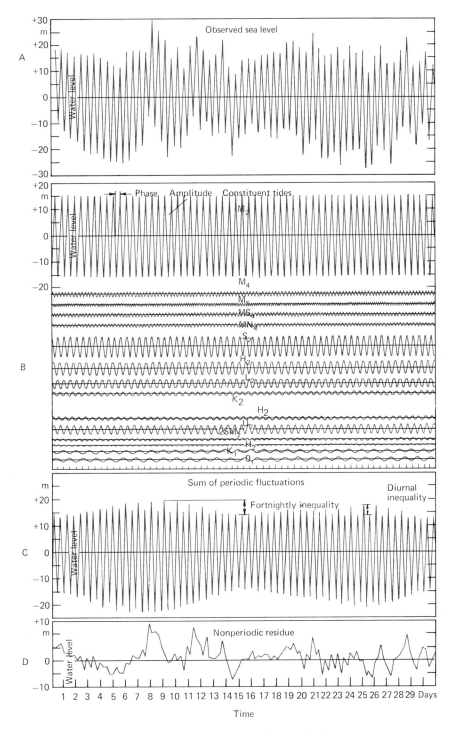

FIGURE 9-14

Harmonic analysis: A) observed fluctuations in water level; B) harmonic constituents; C) sum of harmonic constituents; D) nonperiodic residue due to wind and pressure. (From Defant, 1958, by permission of the University of Michigan Press)

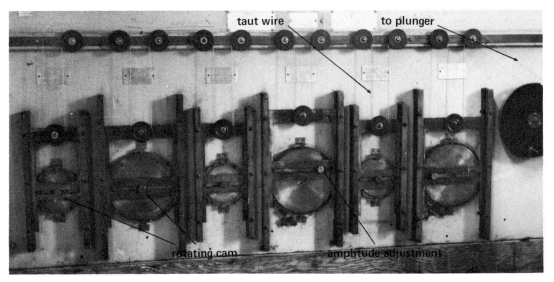

taut wire   to plunger

rotating cam   amplitude adjustment

FIGURE 9-15
Photograph of tidal prediction machine.

are nearly diurnal are rare, but they do occur in such places as Viet Nam, Manila, and parts of the Gulf of Mexico (Fig. 9-1C).

The tides that affect a particular location on a coast are predicted by obtaining a long record of the tide-height fluctuations from the area. This record is analyzed to reveal: (1) the dominant tidal constituents, (2) their amplitude, and (3) their phase (i.e., timing with respect to one another). In other words, the actual record is broken into many partial tides. The partial tides repeat themselves continually, so the tide can be predicted by adding the partial tides expected in the future. Examples of some constants that describe common partial tides are given in Table 9-2. The definitions of the symbols and their periodicities are in Table 9-1. Note, for example, that the M_2 constituent at St. George is one-half the amplitude of that at Pernambuco, whereas the K_1 constituent is twice as great. This fact illustrates the influence that coastal areas have upon the tides.

A tidal curve for Lisbon, Portugal, is obtained by plotting tidal constants from Table 9-2 and then combining the individual curves (Fig. 9-13). A more graphic example is given in Fig. 9-14. Here, an observed tidal record is broken into 15 partial tides. This illustration also shows the influence that meteorological conditions have upon the tide.

The technique just described is incorporated in the operation of a tide-producing machine, a mechanical device used to predict and simulate tides in hydraulic models. The machine shown in Fig. 9-15 simulates the tides in a large hydraulic model of Puget Sound, Washington. It incorporates six cams, each representing a tidal constituent important in that location. Each cam rotates at a rate determined by the periodicity of the tidal constituent it represents. The phase and amplitude of each cam is

150

CORDOVA, ALASKA, 1968

TIMES AND HEIGHTS OF HIGH AND LOW WATERS

JANUARY

DAY	TIME H.M.	HT. FT.	DAY	TIME H.M.	HT. FT.
1 M	0124	12.0	16 TU	0118	11.3
	0642	3.6		0630	4.0
	1248	14.8		1230	13.8
	1930	-2.3		1912	-1.5
2 TU	0206	12.1	17 W	0148	11.7
	0730	3.6		0712	3.6
	1330	14.3		1306	13.8
	2012	-1.7		1948	-1.4
3 W	0248	12.0	18 TH	0218	12.0
	0818	3.6		0754	3.2
	1412	13.4		1348	13.5
	2054	-0.8		2030	-0.9
4 TH	0330	11.8	19 F	0248	12.3
	0912	3.7		0842	2.9
	1500	12.3		1430	12.8
	2136	0.2		2106	-0.2
5 F	0406	11.6	20 SA	0318	12.5
	1000	3.8		0930	2.6
	1548	11.0		1512	11.8
	2218	1.4		2142	0.8
6 SA	0448	11.4	21 SU	0400	12.6
	1054	3.9		1024	2.4
	1642	9.8		1606	10.5
	2254	2.6		2224	2.0
7 SU	0536	11.2	22 M	0448	12.6
	1154	3.9		1130	2.2
	1754	8.7		1724	9.3
	2336	3.7		2312	3.3
8 M	0630	11.1	23 TU	0542	12.5
	1300	3.7		1236	1.9
	1918	8.2		1900	8.6
9 TU	0030	4.7	24 W	0012	4.5
	0718	11.2		0648	12.5
	1412	3.1		1400	1.4
	2042	8.2		2042	8.7
10 W	0136	5.4	25 TH	0130	5.3
	0812	11.3		0800	12.7
	1512	2.4		1518	0.4
	2154	8.6		2206	9.3
11 TH	0242	5.7	26 F	0254	5.4
	0900	11.6		0906	13.0
	1606	1.4		1618	-0.6
	2254	9.2		2306	10.2
12 F	0342	5.6	27 SA	0406	4.9
	0948	12.0		1012	13.5
	1648	0.6		1706	-1.4
	2336	9.8		2354	11.1
13 SA	0430	5.2	28 SU	0500	4.2
	1030	12.6		1112	14.0
	1724	-0.2		1754	-2.0
14 SU	0012	10.4	29 M	0036	11.8
	0512	4.8		0548	3.5
	1112	13.1		1200	14.3
	1806	-0.9		1836	-2.1
15 M	0048	10.9	30 TU	0112	12.3
	0554	4.4		0636	3.0
	1154	13.5		1242	14.3
	1842	-1.3		1912	-1.9
			31 W	0148	12.6
				0718	2.6
				1324	13.9
				1948	-1.3

FEBRUARY

DAY	TIME H.M.	HT. FT.	DAY	TIME H.M.	HT. FT.
1 TH	0218	12.7	16 F	0142	13.3
	0800	2.4		0742	1.0
	1400	13.2		1342	13.6
	2024	-0.5		2000	-0.9
2 F	0248	12.6	17 SA	0212	13.7
	0842	2.3		0824	0.6
	1436	12.2		1418	12.9
	2054	0.5		2036	0.0
3 SA	0318	12.4	18 SU	0248	13.8
	0924	2.5		0912	0.5
	1512	11.0		1506	11.7
	2130	1.6		2112	1.2
4 SU	0348	12.0	19 M	0318	13.6
	1006	2.7		1000	0.6
	1554	9.8		1600	10.3
	2200	2.8		2154	2.6
5 M	0418	11.6	20 TU	0406	13.1
	1100	3.0		1100	1.0
	1648	8.6		1712	8.9
	2236	4.0		2242	4.0
6 TU	0454	11.1	21 W	0500	12.4
	1154	3.2		1212	1.3
	1812	7.7		1900	8.2
	2318	5.1		2348	5.2
7 W	0548	10.7	22 TH	0618	11.8
	1306	3.3		1342	1.3
	2006	7.5		2048	8.4
8 TH	0018	6.0	23 F	0118	5.9
	0700	10.5		0748	11.6
	1430	2.8		1506	0.7
	2142	8.0		2212	9.3
9 F	0154	6.4	24 SA	0300	5.6
	0818	10.7		0912	12.0
	1536	1.9		1612	-0.1
	2242	8.8		2300	10.4
10 SA	0318	6.1	25 SU	0412	4.7
	0924	11.3		1012	12.6
	1624	0.8		1700	-0.9
	2318	9.6		2342	11.3
11 SU	0412	5.4	26 M	0500	3.6
	1018	12.1		1106	13.2
	1706	-0.1		1736	-1.3
	2354	10.4			
12 M	0500	4.5	27 TU	0018	12.1
	1100	12.9		0542	2.6
	1742	-0.9		1154	13.5
				1812	-1.3
13 TU	0024	11.2	28 W	0048	12.6
	0536	3.5		0624	1.8
	1142	13.5		1236	13.5
	1818	-1.5		1848	-1.1
14 W	0048	12.0	29 TH	0112	13.0
	0618	2.6		0700	1.3
	1218	14.0		1306	13.2
	1848	-1.5		1918	-0.6
15 TH	0118	12.7			
	0700	1.7			
	1300	14.0			
	1924	-1.5			

MARCH

DAY	TIME H.M.	HT. FT.	DAY	TIME H.M.	HT. FT.
1 F	0142	13.1	16 SA	0106	14.3
	0736	1.0		0718	-1.2
	1342	12.6		1330	13.2
	1948	0.2		1930	-0.2
2 SA	0206	13.0	17 SU	0136	14.6
	0812	0.9		0806	-1.5
	1412	11.8		1412	12.4
	2018	1.1		2006	0.8
3 SU	0224	12.8	18 M	0212	14.5
	0848	1.1		0854	-1.3
	1448	10.8		1500	11.3
	2048	2.2		2048	2.0
4 M	0248	12.3	19 TU	0248	14.0
	0930	1.5		0942	-0.8
	1518	9.7		1554	9.9
	2112	3.3		2130	3.3
5 TU	0318	11.8	20 W	0336	13.0
	1012	2.0		1042	0.0
	1606	8.6		1712	8.7
	2148	4.3		2224	4.6
6 W	0348	11.1	21 TH	0430	11.9
	1106	2.5		1154	0.8
	1712	7.6		1906	8.2
	2230	5.4		2336	5.6
7 TH	0430	10.4	22 F	0600	10.9
	1212	2.9		1324	1.2
	1936	7.2		2042	8.8
	2330	6.2			
8 F	0548	9.9	23 SA	0130	5.8
	1342	2.8		0748	10.7
	2112	7.8		1448	0.9
				2148	9.7
9 SA	0118	6.6	24 SU	0306	5.0
	0736	10.0		0906	11.1
	1500	2.0		1548	0.3
	2212	8.7		2230	10.7
10 SU	0254	6.0	25 M	0406	3.7
	0854	10.6		1006	11.7
	1554	1.0		1636	-0.1
	2242	9.6		2312	11.5
11 M	0354	4.9	26 TU	0454	2.5
	0948	11.5		1100	12.1
	1636	0.0		1712	-0.3
	2312	10.6		2342	12.2
12 TU	0436	3.5	27 W	0530	1.4
	1042	12.4		1142	12.3
	1712	-0.7		1742	-0.2
	2342	11.7			
13 W	0518	2.1	28 TH	0012	12.7
	1124	13.1		0606	0.6
	1742	-1.2		1218	12.3
				1812	0.1
14 TH	0006	12.7	29 F	0036	13.0
	0600	0.7		0636	0.0
	1206	13.6		1254	12.1
	1818	-1.2		1842	0.3
15 F	0036	13.6	30 SA	0100	13.1
	0636	-0.4		0712	-0.2
	1248	13.6		1324	11.6
	1854	-0.9		1912	1.3
			31 SU	0118	13.0
				0742	-0.2
				1354	11.1
				1936	2.0

TIME MERIDIAN 150° W. 0000 IS MIDNIGHT. 1200 IS NOON.
HEIGHTS ARE RECKONED FROM THE DATUM OF SOUNDINGS ON CHARTS OF THE LOCALITY WHICH IS MEAN LOWER LOW WATER.

FIGURE 9-16

Example of a tide table published by the U.S. Department of Commerce.
The station experiences a mixed tide with a rather large tidal range.

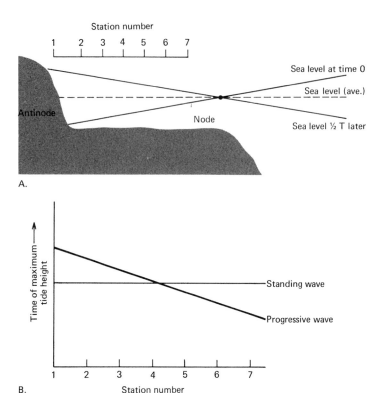

FIGURE 9-17
Schematic of a single-node standing wave in a coastal embayment (A); and the relation between the time of maximum tide height and position in an embayment for progressive and standing waves (B).

adjusted to account for the relative importance of the constituent each represents. All six cams are connected to a large plunger by means of a taut wire. As each cam turns, it displaces the wire and causes the plunger to move vertically. The amount of vertical movement of the plunger is the sum of all cam displacements. The plunger displaces water in the model and causes simulated tides. The model shown in Fig. 9-15 completes a diurnal cycle in approximately 50 seconds.

The tide reference level is different at various places. On the U.S. East Coast *mean low water* represents the *chart datum* to which tide heights are compared. In the Indian Ocean, *indian spring low water* is used; it is the average level of the low-water spring tides. The datum used in Fig. 9-16 is *mean lower low water*, the average of the lower of the two low tides that occur each day.

Scientists formerly did calculations for tide tables on tide-prediction machines. The machine shown in Fig. 9-15 is a small one designed for a model. To predict tides for a general use throughout the world, machines had to be capable of handling over 70 tidal constituents. These machines, once set up, could predict a year's tide in the time of a couple of days. Now, however, large high-speed digital computers have replaced tide-prediction machines.

9.5   TYPES OF TIDES

In a bay, the incoming progressive wave can be reflected at the head of the bay in such a way that the outgoing crest meets the next incoming trough at the mouth.   A node forms at the mouth and an antinode forms at the head (Fig. 9-17A); the water tilts up, then down, in a manner called a *standing wave.*   Every basin of water on earth, by virtue of its size and shape, has the potential of maintaining a standing wave.   Standing-wave tides are produced when the period of natural oscillation of the basin is approximately equal to the tidal period and when frictional effects in the basin do not dampen the oscillation to any great extent.   Resonance causes the tidal range to be amplified to a value limited only by friction. This is the case in the Bay of Fundy.   The length and depth of this estuary are such that the natural period of oscillation is about equal to the semidiurnal tide in the Atlantic Ocean basin.   At the mouth of the bay, there is small vertical movement (3½ m); but at the head of the bay, the maximum tidal range is over 15.4 m.

It is possible to distinguish whether the tide in an inlet or basin is a progressive tide or a standing-wave tide.   A progressive tide rises the same amount but at different times along an inlet.   A standing-wave tide, however, rises different amounts but simultaneously at all points along the inlet (Fig. 9-17B).   Often, tidal behavior in an area is a combination of progressive and standing waves.   Moreover, atmospheric effects such as storms at sea and local winds may tend to modify or obscure the pattern of the tide in any given location.

The natural period of oscillation is important, because it determines the oscillation frequencies that can be sustained in a basin.   Since the Atlantic Ocean basin responds to semidiurnal constituents, diurnal periods are not present.   Conversely, the Pacific Ocean basin responds to both semidiurnal and diurnal periods; hence, mixed tides are common there. This phenomenon occurs to some extent in every irregularity in the world oceans.   It explains why the contribution of each tidal constituent varies from place to place.

9.6   TIDAL CURRENTS

PROGRESSIVE TIDES

The rise and fall of sea level associated with the tides imparts other important motions to the water.   In the open ocean, tidal currents follow paths that approximate the water-particle movements of shallow-water progressive waves.   These paths are highly elliptical or flattened orbits with maximum horizontal velocity under the crest and trough.   Tidal currents of this type are maximum during times of both high and low water and minimum when sea level is midway between these extremes.   Figure 9-18 shows the relationship between water level and tidal current during the passage of a progressive tide.

An estimate of tidal currents can be obtained by calculating the maximum horizontal orbital velocity associated with shallow-water progressive waves if the water depth and tidal amplitude are known. Some calculations of typical cases are shown in Table 9-3.

Table 9-3 indicates that tidal currents are swifter in shallower water. Tidal amplitudes are larger in shallower water, so a 62 cm/sec estimate for currents along the continental shelf (100 m depth) is approximately correct. Tidal current velocities are the same from the sea surface to near the bottom where friction retards the flow.

STANDING-WAVE TIDES

The sequence of tidal currents associated with standing-wave tides is illustrated in Fig. 9-18B. Notice that maximum current speeds occur at mean tide level and that current speeds are nearly zero at maximum flood and ebb, just the opposite of the case for progressive tides. If the tides are semidiurnal, maximum current occurs approximately 3 hours before high or low water.

FIGURE 9-18
Relation between tide height and current speed for a progressive wave tide (A); and a standing wave tide (B).

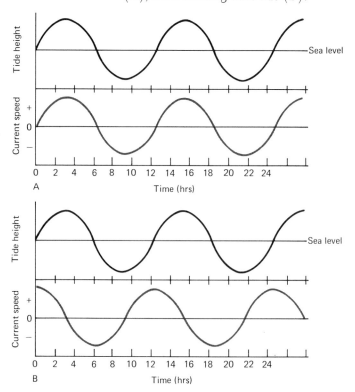

TABLE 9-3

ESTIMATES OF
MAXIMUM TIDAL CURRENTS OF SEMIDIURNAL TIDES
(CALCULATED FROM SHALLOW-WATER PROGRESSIVE-WAVE
ORBITAL VELOCITIES)

Depth	Tidal amplitude of 1 m	Tidal amplitude of 2 m
100 m	31.3 cm/sec*	62.6 cm/sec
500 m	14.0 cm/sec	28.0 cm/sec
2,000 m	7.0 cm/sec	14.0 cm/sec
4,000 m	4.9 cm/sec	9.8 cm/sec

* 50 cm/sec = about 1 knot = about 1 nautical mile per hour

ROTATIONAL EFFECTS

The rotation of the earth causes an apparent deflection of tidal currents. In the northern hemisphere, the deflection is to the right, or clockwise. In the southern hemisphere, it is counterclockwise. These rotational motions are developed best in approximately equidimensional basins of the major oceans. Figure 9-19A shows the rotary systems thought to exist in the Atlantic Ocean basin. The cotidal lines on this figure represent the geographical positions of the tidal crest at hourly intervals over a 12-hour semidiurnal period. The point where cotidal lines merge is an *amphidromic point*, a place where no tidal fluctuation of sea level exists. The tidal range is zero at an amphidromic point; it increases with distance from that point in an approximately regular manner. Figure 9-19B illustrates the motion of the water surface in an amphidromic tidal system.

The positions of amphidromic points and cotidal lines in the deep sea are inferred from tidal measurements made at nearshore sites on continents and oceanic islands. The results of the measurements are then extrapolated into the deep sea. Direct measurement in the deep sea would greatly improve the oceanographer's understanding of oceanic tides.

NEARSHORE TIDES

Tides in nearshore regions greatly differ from those assumed to exist in the deep sea. Because of the diverse shapes and sizes of bays, sounds, and straits, tides and tidal currents vary greatly. The strongest tidal currents exist where the tide must enter a bay through a constricted entrance. If the bay behind the constriction is large and has a significant tidal range,

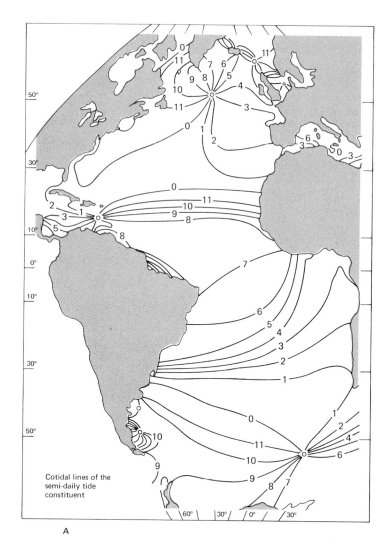

Cotidal lines of the
semi-daily tide
constituent

A

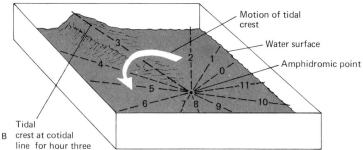

Motion of tidal
crest

Water surface

Amphidromic point

Tidal
crest at cotidal
line for hour three

B

FIGURE 9-19

(A) Cotidal lines of the semi-daily tide constituent of the Atlantic Ocean
Basin; (B) Motion of the water surface in a schematic amphidromic tidal
system in the northern hemisphere.

then the flood and ebb currents through the entrance must be large. San Francisco Bay is such an example; tidal currents of 300 cm per sec have been measured during spring tides.

Tidal currents are also high where channels connect different bodies of water. Strong hydraulic currents arise if the water-level fluctuations in the two basins are out of phase. One example is Akutan Pass in the Aleutian Islands where tidal currents of 450 cm per sec have been recorded. Currents exceeding 300 cm per sec have been measured in the Bungo Strait entrance to the Sea of Japan.

TIDAL BORES

In some tidal rivers, the incoming tidal wave becomes steepened into a wall of turbulent water called a *tidal bore*. The exact cause of tidal bores is not known. However, they have a critical relationship to the characteristics of the incoming tidal wave, slope and shape of the channel, mean channel depth, and river flow. Few rivers develop well-defined tidal bores, although spectacular examples of bores exist. For instance, the Tsientang River bore in mainland China is 3.7 m high and surges upriver at 800 cm per sec. The bore of the Amazon River reaches 5 m and rushes upstream at 600 cm per sec for a distance of 480 km. Figure 9-20 shows the tidal bore of the Severn River in England.

FIGURE 9-20
The Severn bore rushing up the river at Stonehead, Gloucester, England.

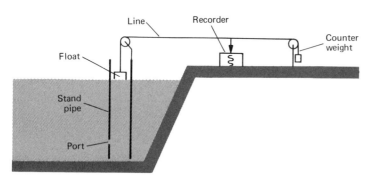

FIGURE 9-21
A simple tide recorder.

9.7 MEASURING THE TIDES

It was mentioned in Section 9.4 that, in order to predict the tides in a given place, they must first be measured and analyzed for certain tidal constants. For this reason, tidal stations are located in thousands of coastal areas around the world. There are many types of tidal measuring devices; in this section, one of the more simple types will be discussed. Another type is mentioned in the discussion of oceanographic instruments in Chapter 15.

A self-registering tide gauge is one that can be set in a coastal region to operate continuously with minimal upkeep or maintenance. The gauge is located on a pier or sea wall. It consists of a tube or reservoir that has a very small opening to the sea (Fig. 9-21). The size of the opening relative to the cross section of the tube is critical in order to restrict the flow of water. If a surface wave with a 4-second period passes, the flow of water through the orifice during that time must be so slight that the level in the reservoir does not change appreciably. In other words, the restricted opening acts as a filter that eliminates shorter period waves from the tide record. Inside the reservoir there is a float that is connected mechanically to a stylus. The stylus places a continuous mark on a turning drum, thus making a record of the tidal variations.

9.8 THE TIDES AS A SOURCE OF POWER

Under the right conditions it is possible to use the tides as a means of generating electrical power. For this application, the tidal range must be greater than 3 m, and the water must flow into a rather large basin through a narrow constriction. Because of these constraints, there are relatively few areas in the world that can support a tidal generating plant. Some of these are the Rance estuary off the coast of Brittany, Passamaquoddy Bay in the Bay of Fundy, the San Jose and Deseado rivers in Argentina, Cook Inlet in Alaska, the Severn estuary in England, and Penzhinskaya Bay in Siberia.

The tidal generating facility of the French government in the Rance estuary exploits a maximum tidal range of 11.3 m. The basin into which the water flows has an area of 20 sq km. A dam 710 m long has been built across the entrance to the bay and two-way turbines have been placed so that power is generated on both ebb and flood tides. The facility has been in partial operation since 1960 and will ultimately produce 565 million kw hr annually. This amount is equivalent to the energy produced by the combustion of ½ million tons of coal. It could support the electrical needs of an industrialized city with a population of 75,000 persons.

READING LIST

CHRISTOPHER, P., H. RUSSEL, AND D.H. MACMILLAN, *Waves and Tides*. London: Hutchinson's Scientific & Technical Publications, 1952. 348p.

DARWIN, G.H., *The Tides and Kindred Phenomena in the Solar System*. San Francisco, Calif.: W.H. Freeman and Co., 1962. 378p.

DEFANT, A., *Ebb and Flow: The Tides of Earth, Air and Water*. Ann Arbor: The University of Michigan Press, 1960. 121p.

NICHOLSON, T.D., "The Tides," *Natural History*, LXVIII, No. 6, June–July 1959, 326–333.

WARBURG, H.E., *Tides and Tidal Streams*. New York: Cambridge University Press, 1922. 95p.

ONE
TWO
THREE
FOUR
FIVE
SIX
SEVEN
EIGHT
NINE
TEN
ELEVEN
TWELVE
THIRTEEN
FOURTEEN
FIFTEEN

inshore oceanography

At the coastal zone of the world ocean, air, water, and solid earth meet; and there is endless interaction among the geological, biological, meteorological, and oceanic processes. Each of these environmental processes affects the nature of a coastal region to some degree. Consequently, a study of the processes at work in a coastal sector can help us understand the origin, relative age, and history of that coast.

10.1 THE OPEN COAST

OCEANIC INFLUENCE

The diverse influences of ocean waters upon coastal features are related to both the physical attributes of seawater (i.e., waves, currents, and turbulence) and the chemical properties (i.e., solubility and concentration). The effect of ocean waves, however, is the most important. The configuration of a coastal area and the offshore floor of the sea is largely the result of wave action. For this reason, the properties of shoaling waves are emphasized in this section.

NEARSHORE CIRCULATION. As waves carry energy toward shore, they encounter shallow water and their speed of propagation decreases. If the

Figure 10-1
A large ocean swell refracts against
a complex coastline. (Photograph
courtesy of Barbee Scheibner).

A

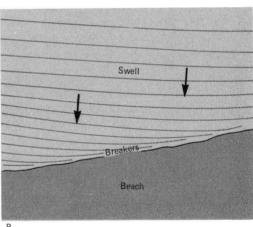

B

Figure 10-2

Photo (A) and line drawing (B) of wave refraction just seaward of the surf
zone. The direction of longshore currents is from right to left along the beach.

direction of wave attack is not perpendicular to the shoreline, the shoreward part of a wave slows more than the seaward portion and *refraction*
occurs (Fig. 10-1). Figure 10-2 shows how the process of refraction
causes wave crests to align themselves parallel to the shoreline. However,
this alignment is only a tendency. Wave crests seldom become completely parallel to the shore except where the beach has a long, gentle
offshore profile.

161

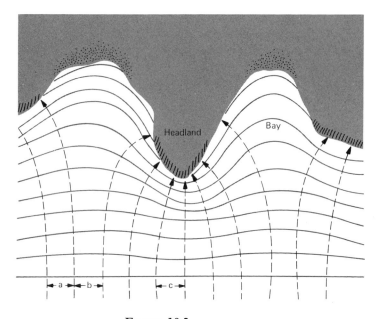

FIGURE 10-3
Schematic of refraction on an irregular coast. Waves are approaching parallel
to the coastal trend. Solid lines represent wave crests. Dashed lines
are wave orthogonals. (After A. N. Strahler, 1963).

The behavior of waves in expending their energy on a beach can be
illustrated by the use of *orthogonals*. These are imaginary lines drawn in
such a way that they divide the crest of an unrefracted wave into equal
segments of length and energy. Orthogonals are always perpendicular to
wave crests and thus show the direction of wave propagation (Fig. 10-3).
At the same time, orthogonals indicate the distribution of energy in a wave
train; the amount of energy contained between any pair of orthogonals is
assumed to be constant regardless of refraction. Where waves attack an

FIGURE 10-4
Direction of littoral currents resulting from waves breaking at an
angle to the shore.

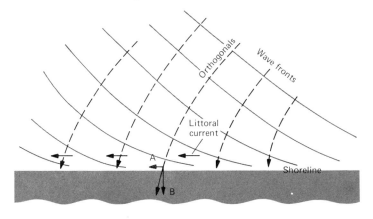

irregular coastline, a point of land refracts the waves so that energy is focused on it; an embayment has the opposite effect. In Fig. 10-3, the amount of energy moving onshore is the same for sections a, b, and c. However, this energy is spread over a longer stretch of beach at B and a shorter stretch of beach at c; consequently, the energy per unit length of coastline is increased in section c and decreased in section b. The net effect is that wave erosion is greater on the headlands. Given enough time, it tends to straighten an irregular coastline, regardless of the direction of wave attack.

As a wave approaches shore, the orbital velocity at the wave crest continually increases, whereas the propagation speed decreases. Eventually, the crest "overruns" the trough, and the wave breaks. Theoretical considerations as well as field observations indicate that breaking occurs when either of two conditions are met: (1) the water depth becomes less than 1.28 times the wave height, or (2) the wave steepness (H/L) surpasses 1/7. After a wave breaks, the water-particle motion becomes intensely turbulent, and energy approaches the beach in a wave of translation rather than oscillation.* The mass of turbulent water that moves upon the beach is called *swash*. The water that runs back down the beach under the influence of gravity is termed *backwash*. Backwash removes the water from the beach and causes it to oppose the oncoming waves. In so doing, the oncoming waves are steepened and break a bit sooner.

The transport of water inside the breaker zone is in the direction the waves were traveling just prior to breaking. In Fig. 10-4, the transport is resolved to show a component onto the beach (B) and a component along the beach (A). The upbeach transport (swash) is balanced by backwash, whereas the transport of water along the beach forms a *longshore*, or *littoral*, *current*. The quantity of water transported by a littoral current is related to the character of the approaching waves and the angle of approach. The larger the waves or the greater the angle of approach, the stronger the longshore current.

On an irregular shoreline, the net movement of water is from headlands toward embayments (Fig. 10-5). As a result, there is a *convergence* of longshore currents at the mouth of a bay where water accumulates. This accumulation causes a narrow, swift current, called a *rip current*, which moves seaward from the convergence zone (Fig. 10-6).

This example shows only one way in which rip currents may originate. Actually, they are associated with any situation where water accumulates in the surf zone to a point where the excess water flows seaward. Several factors can cause such accumulation: converging longshore cur-

* Depending on the way in which breaking occurs, the oscillatory characteristics of a wave are often maintained to the extent that the wave may reform and break several times as it approaches the beach. Each succeeding breaker is smaller than the previous one.

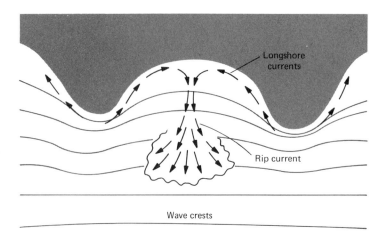

Longshore currents

Rip current

Wave crests

FIGURE 10-5
Net movement of water in an embayment cause rip currents to form.

rents; a physical obstruction to longshore currents, such as a rock groin (a barrier, illustrated in Fig. 10-18), breakwater, or headland; or just the continuous addition of water from breakers approaching parallel to the beach. Rip currents can be recognized in several ways: discoloration due to suspended sand, premature steepening of the approaching waves and the seaward displacement of the breaker line, or the accumulation of foam at the head of the rip. A small cusp of sand may also occur on the beach where the convergence or obstruction occurs.

FIGURE 10-6
Photograph of rip currents in an embayment.

Some rip currents extend as far as 1000 m offshore, are 30 m in width where they flow through the surf, and travel up to 100 cm per sec. These currents carry sand as well as excess water from the beach. Because of their velocity, they often erode shallow channels through the surf zone.

The well-developed rip-current system exhibits a circulation pattern as shown in Fig. 10-7. Rip currents may be semipermanent features or may last only a few hours or a few days. They are probably the greatest single cause of drownings on a beach. A swimmer caught in a rip current (also called an *undertow*) can escape being swept out to sea if he swims parallel to the shore rather than trying to swim against the direction of the current; the latter recourse only enhances the possibility of exhaustion.

To summarize, the effect of waves approaching an irregular coastline is the formation of circulation patterns related to the direction of wave approach and the shape of the coastline. Water is pushed onshore by the waves and flows offshore as rip currents. Littoral currents occur within the surf zone and move from areas of relatively high energy (headlands) to low energy (embayments).

NEARSHORE SEDIMENT TRANSPORT. The nearshore circulation system is very effective in moving sand. Waves provide the energy to erode the coastline and temporarily suspend sedimentary particles, while the longshore currents transport the sediment along the coast. The process of suspension and movement of sediment in the surf zone is called *longshore*, or *littoral, drift*. Laboratory and field studies have shown empirically that the sand transported as littoral drift can be related to the wave energy expended on the beach.

Moving water transports sand according to its *competency*, or the

FIGURE 10-7

Schematic diagram of the nearshore circulation pattern on a straight beach. (From F. P. Shepard, Submarine Geology, 2nd edition, Harper & Row, New York, 1963).

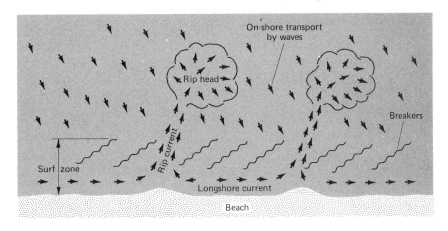

A

FIGURE 10-8
A steep rocky beach (A) and flat sandy beach (B) illustrate the association of beach slope with sediment size.   (B courtesy of Clifford E. Moon).

B

FIGURE 10-9
The sediment size in an embayment decreases from rocks on the
exposed headland where wave energy is high to sand at the back
of the bay where energy is low.

size of material that can be moved by a given current, and according to its
capacity, which is the quantity of material that the flow is capable of
moving. With increasing current speed, both competency and capacity
increase. Hence, large waves are usually associated with higher rates of
littoral drift and move coarser sediment than small waves do.

In general, beaches on stormy coasts (e.g., in high temperate latitudes or polar regions) are steep, narrow, and consist of coarse sand, peb-
bles, or cobbles (Fig. 10-8A). Protected beaches and those located in
many tropical areas, where winds and waves have low energy, frequently
are broad, have gentle slopes, and are composed of medium to fine sand
(Fig. 10-8B). Similarly, around headlands, wave energy is high; in embayments, waves are smaller. Bays, therefore, have sandy beaches, whereas
the coast along headlands is rocky or consists of quite coarse material
(Fig. 10-9).

In the swash-backwash zone, sand is transported up the beach face
with the swash and moves directly downslope during backwash. Figure
10-10 shows how this motion results in a net drifting of sand along the
beach. This *beach drift* is generally parallel to the direction of longshore
drift but is highly irregular. Note that the direction of these two modes
of sediment movement is dependent upon the direction of wave attack. If
the direction of wave attack is resolved into components parallel and perpendicular to the direction of the beach, the direction of the parallel com-

167

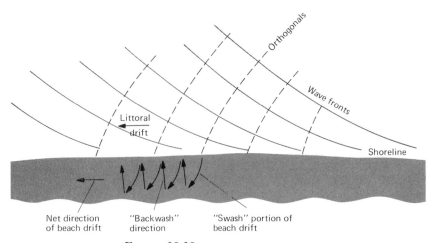

FIGURE 10-10
Sediment movement along the beach occurs in the surf zone as littoral
drift and on the beach face as beach drift.

ponent indicates the direction of sediment drift (see Fig. 10-4), and its
length indicates the relative strength of the current.

The two processes, beach drift and littoral drift, transport beach
materials from river mouths and eroding headlands to their ultimate desti-
nation in embayments or, via submarine valleys and canyons that intercept
littoral drift, to the sea floor lying below depths affected by wave activity.

The sands, gravels, and coarser materials on the beaches have formed
gradually throughout geologic time. Indeed, some sands have existed as
such for centuries. In other cases, sand has gone through several cycles of
incorporation in sedimentary rocks (sandstones), erosion, transportation to
the beach, and movement in the littoral zone. In a geological sense, involvement in such cycles is transitory, as are the beaches we enjoy today.
Compared to the quantity of sand that has been made throughout geologic
time, the amounts being introduced today are relatively small. Therefore,
the net removal of sand to below the limit of wave activity must proceed
at a relatively small rate; otherwise, the coasts along the world ocean
would be barren rock cliffs and wave-cut platforms except near the mouths
of large rivers.

THE BEACH PROFILE. Direct observations of beaches establish that
sand transport in the littoral zone is primarily along the shore. Where
prevailing winds and semipermanent storm centers cause waves to attack
the coast chiefly from one direction, there is a natural tendency for the
beach to adjust itself to the average waves and the rate of sediment supply.
The distribution of wave energy will cause the beach either to *prograde*
(build out) locally or to recede so that a dynamic equilibrium is eventually
attained. This equilibrium often reflects some sort of annual average wave
condition. However, the sediment in a sector can vary from being eroded

168

in winter to being prograded during the summer, because the nature of the waves changes seasonally. Figure 10-11A shows a beach profile as it changes throughout the year. From April to September, the beach is prograding, whereas it retreats during the winter season (September to March).

Laboratory studies and measurements on beaches show that short-period, steep waves from local winter storms occurring close to shore have an increased capacity and tend to keep sand suspended. Under these conditions, sand is removed from the beach by rip currents and deposited in the seaward part of the littoral zone. During the summer, the beach is attacked by swells with long wavelengths and low steepness that arrive from distant storms. These waves influence the floor of the sea at great distances offshore and gradually return sand shoreward to repair the beach erosion of the previous winter. A winter profile is recognizable by a steep beach face, relatively coarse sediment, and a sea cliff or winter berm (a narrow shelf). A summer beach profile is characterized by finer sediments, less steep beach face, and smaller berm. Sometimes certain portions of

FIGURE 10-11

The beach configuration changes throughout the year as a function of wave conditions (A). The characteristic elements of a beach profile (B). (A after Office of Naval Research, 1957; B after A. N. Strahler, 1963).

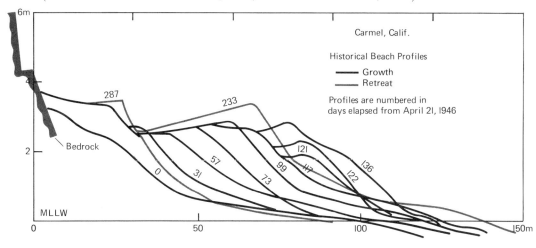

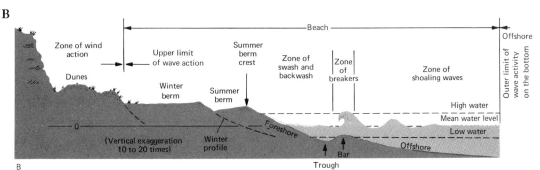

the winter profile (i.e., the winter berm or some offshore bars) remain as remnant features because of the inability of the summer waves to affect them (Fig. 10-11B).

We can often determine the state of the sediment and the short-term processes acting on a beach sector by examining certain features of a beach. Its configuration in profile and plan, distribution by grain size of beach sediments, and types of vegetation at the shore indicate whether a beach is being eroded or accreted.

In addition, a beach's features reveal the types of waves acting on it. The slope of the beach and the height of the berm are determined by the energy in the swash of the waves. An energetic swash will carry sand to the top of the beach berm and deposit it there, thus steepening the face of the beach. If the energy of the waves increases suddenly, a sea cliff quickly forms because of the increased rate of erosion. Under prolonged attack, however, a smooth slope eventually results. If waves are so energetic that the swash overtops the berm, the beach will build upward in response. If this action occurs infrequently, coastal winds may erode the berm and build a dune field behind the beach. When conditions remain relatively constant for several years, certain types of vegetation, crawling plants, and beach grasses (Fig. 10-12) will become established and stabilize the beach in that configuration.

The distribution of various types of sand on a beach reflect the prevailing regimen of waves. Extremely coarse and heavy materials tend to

Figure 10-12
Grasses growing on the beach face help to stabilize the sand
and promote further deposition.

A

B

FIGURE 10-13
Sedimentary particles are sorted
by wave action according to size
and density, producing layers or
bands paralleling the beach.
Examples of sorting are seen by
bands of shell fragments (A), or
layering of dark high density
minerals (B). The dark layers in
(B) are approximately 1–3 mm
thick.

remain where the strongest waves deposit them, usually along the highest
berm. Extremely fine and light materials tend to persist in suspension
and so are removed downcoast and eventually seaward. Variations in the
normal wave patterns are shown vividly where dark, heavy minerals are
present in minor quantities in a light-colored sand. The heavy minerals
remain after the differential erosion of the light sand; in this way, lenses
and stringers of heavy materials become segregated. Sometimes layers or
zones of concentrated shell fragments are separated by the same process

171

FIGURE 10-14
Coastal sand dunes represent a depository of excess beach sand.

(Fig. 10-13). These layers often are exposed in sea cliffs that form when steeper waves begin to attack the shore.

NEARSHORE DEPOSITION AND EROSION. Scientists do not completely understand how sand movement onshore and offshore responds to wave conditions. Much of what we know about this complex mechanism comes from studies of beach models in wave tanks where scaled experiments are conducted. Although natural beaches do not always corroborate observations of beach behavior in models, at least some broad generalizations can be drawn. Empirical measurements on natural beaches are generally not satisfactory, because the generalizations derived usually apply only to a single location.

The movement of sand along the coast is the result of many individual littoral drifts acting on discrete coastal segments. Locally, the currents move sand from promontories toward embayments. Regionally, the drift tendency is in the direction of the longshore component due to the prevailing wind waves. Most littoral sand is introduced at the mouths of streams. Sand is lost as it is carried by steep, high waves into deep water or submarine canyons, or as it is removed to dunes by onshore winds (if rainfall is not excessive) as seen in Fig. 10-14. The long-term effect of all these factors is a straightening of an irregular coastline.

Whenever either a longshore current is interrupted or the waves that provide the energy to maintain the longshore current are interrupted, the capacity of that flow to transport sediment is decreased and deposition occurs. Figure 10-15 shows how an embayment interrupting a straight

172

coast alters the wave refraction in such a way that deposition of littoral drift occurs, causing the formation of a spit or small point of sand across the mouth of the bay (Fig. 10-16). In cases where the embayment has a vigorously flowing river at its head, the estuary will never become completely sealed. However, the river may be forced to migrate because of fluctuation in the dynamic equilibrium between sediment deposition from littoral drift and sediment erosion by outflowing river water (Fig. 10-17). Often the way in which a spit has been built indicates the direction of sediment drift that has persisted in that area for a considerable number of years. If we extend this reasoning, we can sometimes infer the direction of the prevailing winds or waves.

Frequently, the natural tendency of littoral processes is not de-

FIGURE 10-15

Formation of a spit across an estuary. The spit builds into the estuary because of wave refraction into the mouth of the bay. (A) initial situation; (B) well-developed spit.

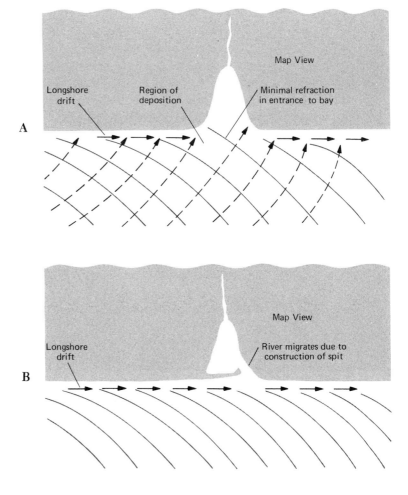

FIGURE 10-16
The straight narrow spit forming this tidal embayment was built by longshore drift moving from right to left. (Photograph courtesy of Dennis Byrne).

FIGURE 10-17
Due to longshore drift a spit has been built across a river mouth. Note how the river has been forced to erode the opposite bank as a result of spit growth. (Photograph courtesy of Peter B. Taylor).

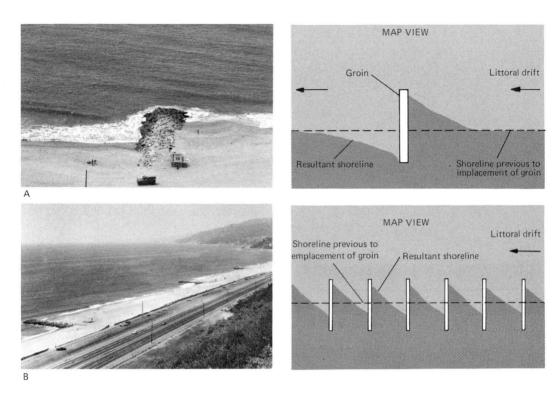

FIGURE 10-18
Photograph and line drawing of (A) nearshore refraction and sedimentation around a single groin; (B) sedimentation associated with a series of groins.

sirable. Coastal installations are jeopardized if excessive erosion occurs where they are built or if they are built on an unstable beach, in which wide variations in accretion and erosion take place yearly or over a few years. Sometimes a man-made interruption of littoral drift precipitates sudden changes in the configuration of the coastline. In all these cases, it is desirable to stabilize the beach and prevent excessive erosion or accretion. In other words, littoral drift must be stabilized. In practice, it is erosion of beaches that must be controlled, for losses of sand during a prolonged period of erosion tend to be permanent losses.

Techniques for controlling beach erosion involve such procedures as retarding the littoral drift rate, trapping sand or otherwise diverting sand from the heads of submarine canyons, building jetties and breakwaters to protect segments of beach exposed to concentrated wave attack, or bringing in sand to nourish eroding beaches.

Figure 10-18A illustrates the effects of a deliberate interruption of sediment drift. This figure shows a groin, or barrier, constructed to extend from the beach out into the breaker zone. To be effective, this groin must extend from the bottom of the surf zone to high-tide level. When a groin is installed, it dams the littoral drift. Thus, sand accumulates as sedi-

175

mentation on its updrift side; subsequently, sand is removed (erosion) on its downdrift side. Figure 10-18B shows the effect of a series of groins constructed to increase the area of a sand beach.

Deposition of littoral drift also occurs when the waves that provide the energy to move sand are interrupted. A simple breakwater parallel to shore may decrease wave surge in a harbor entrance, but it also causes sand deposition as shown in Fig. 10-19. Some other examples of common breakwater configurations are also shown. Note that both a zone of accretion and a zone of erosion are associated with the breakwater. Generally, sand is dredged or pumped from the site of deposition to the zone of erosion. Although the cost is considerable, this procedure maintains the harbor entrance and minimizes damage by erosion.

FLUCTUATIONS IN SEA LEVEL

Sea-level changes have a considerable influence on coastal processes and on beaches. A large tidal range is often associated with broad beaches and extensive salt marshes. Wave energy is greatly diminished by friction over the broad intertidal platform, and coastal erosion is intermittent. The strength of tidal currents is associated with tidal range. In coastal areas exposed to both large waves and a large tidal range, erosive processes can be significant in the nearshore zone.

Changes in sea level have occurred in past geologic time as a result of vertical movements of the land and changes in the volume of water in the ocean basins. The land can move either up or down and submerge or raise coastal areas. Such movements can be related to faulting, folding, and tilting of the earth or to isostatic adjustments of the crust when heavy loads of ice, sediment, or lava are added or removed. For example, crustal subsidence generally occurs in the vicinity of large deltas; crustal rebound follows a period of glaciation.

Coasts that have been influenced by recent earth movements are recognized in New Zealand, New Guinea, Japan, California, and around the Mediterranean Sea. Isostatic movements from the melting of continental glaciers are recognized in Scandinavia and Canada.

The continental glaciers of the Pleistocene epoch removed significant volumes of water from the ocean basins. Geological evidence from the land has led scientists to conclude that the Pleistocene epoch was characterized by alternate advances and retreats of the ice of continental glaciers. Sea-level fluctuations accompanied these oscillations in ice formation.

Calculations of the volume of water frozen in glaciers during the last glacial stage indicate that the sea level must have been lowered by 110 to 130 m. These calculations are corroborated by fathograms of submerged shorelines and by beach sands dredged from depths to 130 m throughout the world.

The ice of the last glacial stage began to melt about 20,000 years

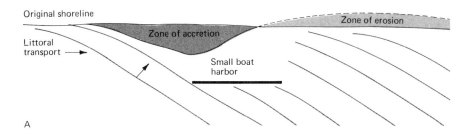

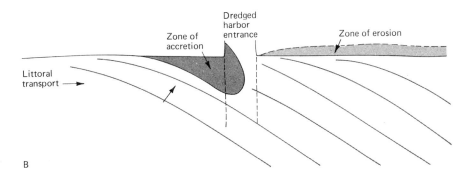

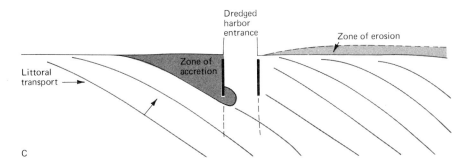

FIGURE 10-19
Deposition of sediment associated with the interruption
of longshore currents in the nearshore zone.

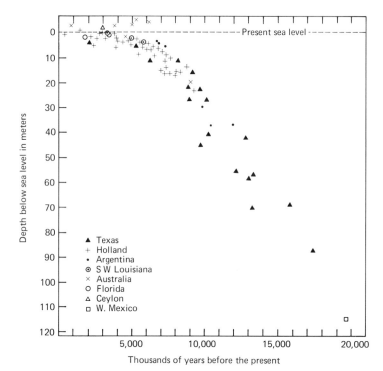

FIGURE 10-20

Estimates of the Holocene marine transgression during the past
20,000 years. (Permission of F. P. Shepard).

ago. Since then, the sea level has risen approximately 120 meters (Fig.
10-20). All of the estuaries, embayments, and coastal regions of the world
have also been inundated by the Holocene marine transgression of the
past 20,000 years. Most coastal features represent effects of erosion, deposition, and biological growth that have occurred in the past few thousand
years.

GEOLOGIC INFLUENCE

The degree of erosion in a coastal zone depends on the differences
in hardness and durability of rocks. Variations in the character or structure
of coastal rocks exposed to waves and seawater thus are accentuated by
differential physical or chemical erosion.

The resistance of a rock can be measured by (1) its hardness or
ability to withstand constant wave attack, (2) its chemical resistance to
the solvent properties of seawater, and (3) its ability to undergo processes
of continuous wetting and drying. On exposed coasts where wave forces
are great, the harder rocks stand as rocky promontories after the softer ones
have been removed (Fig. 10-21), or they form bold cliffs surrounded by
shore platforms. The durability of rocks exposed to identical oceanic conditions can be compared in Figs. 10-22 and 10-23.

Figure 10-21
Resistant rock formations may control the erosion patterns and thus the coastal configuration. Here, more resistant strata remains as stacks and headlands after the softer rock has been eroded by wave action. (Photograph courtesy of Joe S. Creager).

Figure 10-22
Numerous offshore stacks give evidence of rapid retreat of the coast due to marine erosion.

Figure 10-23
When rock formations are more durable, cliff erosion proceeds more slowly. Here, the horizontal flow planes of the lava flows control the orientation of the shore platform.

Figure 10-24
The orientation of bedding planes in coastal formations can influence the erosion patterns. Here, the dipping strata cause a serrated shore platform and beach cliff. Average thickness of layers is approximately 15 cm. (Photograph courtesy of Joe S. Creager).

Figure 10-25
Aerial view of a fringing reef shows how coral growth extends the coast and supplies sand for adjacent beaches. (Photograph courtesy of Joe S. Creager).

These figures show adjacent coastal regions on the Mornington peninsula near Melbourne, Australia. One coast is composed of an ancient dune field that has been cemented with calcium carbonate. These rocks are relatively weak, and stacks extend a distance offshore, indicating a kilometer of coastline retreat since the Holocene marine transgression. Several miles away, the coastal rocks are massive and durable basalts. These basalts are exposed to the same waves as the sandstone, but cliff erosion is only about one-tenth as much. Note that the horizontal bedding planes between successive lava flows in Fig. 10-23 also control the level and inclination (in this case, horizontal) of the shore platform. Gently dipping coastal formations often impress this "signature" on the short platform. Figure 10-24 shows an example of inclined slopes and benches of the shore platform. Also evident are the effects of differences in the resistance to erosion of alternating layers of rocks.

180

BIOLOGIC INFLUENCE

Certain organisms living in the coastal zone likewise influence the genesis, shape, and structure of coastal landforms. Noteworthy examples are the colonies of reef-building calcareous organisms on exposed coasts and the marsh grasses and mangroves in protected waters.

Reef corals and calcareous algae flourish in tropical regions of the world (as described in Chapter 2). These organisms produce massive platforms that can extend seaward for 130 km. Buccoo Reef on Tobago Island in the West Indies is an example of a fringing reef that not only actively extends the coast seaward but also provides carbonate sand for the island's beaches (Fig. 10-25).

Coastal erosion can similarly be caused by biological processes. Some marine organisms living in the surf zone can dissolve or abrade the rock mass to which they cling. Certain mollusks and echinoderms actually carve holes in the rock to provide a shelter from waves. In some areas, the feeding activity of snails wears away the sea cliffs (Fig. 10-26). In fact, scientists generally agree that, under favorable conditions, the biochemical degradation of limestone in the shore zone surpasses the effects of purely physical and chemical processes.

Coastal vegetation plays a primary role in both stabilizing and

FIGURE 10-26

In browsing the rocks for alga, snails carve small pits into this sandstone which accounts for significant coastal erosion.

FIGURE 10-27
Various forms of salt tolerant grasses cause deposition of sediment and thus extend protected coasts seaward. (Photograph courtesy of Larry Lewis).

building up the shore environment. Several forms of grass grow on sand dunes and cliffs exposed to wind and salt spray (Fig. 10-12). These grasses strengthen the beach's resistance to erosion and cause further deposition by trapping windblown sand. Grasses have been introduced into coastal areas for the specific purpose of stabilizing drifting coastal sand.

For example, marine grasses like Zostera (eel grass) act as stabilizers of subtidal mud and sand flats. Other salt-tolerant plants colonize the margins of protected marine waters. There are pioneer colonizers like salicornia. In tropical estuaries, we find the mangrove. All these plants inhabit intertidal mud banks and cause the drainage to follow definite channels. Thus, they promote deposition of mud by decreasing the speed of currents in certain areas. Inevitably, accretion of mud follows the spread of these plants into a bay or lagoon (Fig. 10-27).

10.2 THE ESTUARINE ENVIRONMENT

Men have always lived closed to estuaries, that is, at the mouths of rivers, because these bodies of water provide large quantities of food and give shelter to industrial activity and commerce. Furthermore, estuaries used to provide an easy means of waste disposal. It has been only recently that men have realized the vulnerability of river mouths and, indeed, the entire watershed of a river.

Prior to the mid-19th century, abuses of the estuarine environment

182

were limited to silt erosion from agriculture, overgrazing, and deforestation. Since the industrial revolution, cities, factories, and great transportation facilities have centered around the protection and water afforded by coastal embayments. Today, seven out of the ten largest metropolitan areas of the world have grown up around estuaries. It has been estimated that one-third of the population of the United States lives or works adjacent to its major estuaries. As a result, many coastal bodies of water are suffering from the contaminating effects of industrial and human waste products. An understanding of estuarine processes, therefore, is necessary and critical if men are to use these natural bodies of water without ruining them to the point of irreversible damage.

TYPES OF ESTUARIES

An estuary can be defined as a "semi-enclosed coastal body of water which has a free connection with the open sea and within which sea water is measurably diluted with fresh water derived from land drainage." From a geomorphological standpoint, estuaries can be divided into four basic types. These are: (1) coastal plain estuaries, or drowned river valleys, (2) fjords, (3) bar-built estuaries, and (4) estuaries produced by tectonic processes. Each of these types has certain attributes associated with its geography, geometry, catchment basin, bathymetry, and offshore oceanic characteristics.

COASTAL PLAIN ESTUARIES. The rise in sea level at the end of the Pleistocene epoch caused extensive flooding of lowland areas throughout the world. On broad coastal plains, seawater extended up Pleistocene river valleys to form what are called *coastal plain estuaries*. The shoreline of such drowned river valleys follows the drainage pattern of the lower reaches of the river (Fig. 10-28).

FIGURE 10-28
The rising sea level of the Holocene marine transgression caused seawater to fill the lower portions of existing river valleys.

The circulation within a coastal plain estuary is related to the river flow and the tidal currents. Seawater flows inward along the bottom while the river flows outward at the surface. Mixing occurs in the region of density change between the two layers (Fig. 10-29). The extent of salt-water intrusion is controlled by the tide. The Mississippi River is a prominent example of a *salt-wedge estuary*, in which the river flow is so strong that it obscures the tidal influence on the incursion of seawater along the bottom. The river flow seaward and the seawater intrusion at the river bottom maintain a salinity distribution in the estuary, so that a

FIGURE 10-29
Representation of longitudinal profiles of coastal plain estuaries showing the salt wedge type (A); and the partially mixed type (B).

A Salt wedge estuary

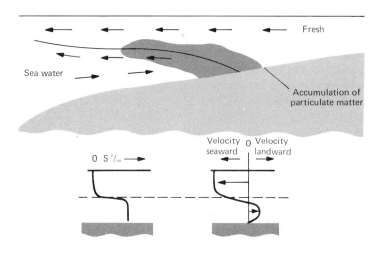

B Partially mixed estuary

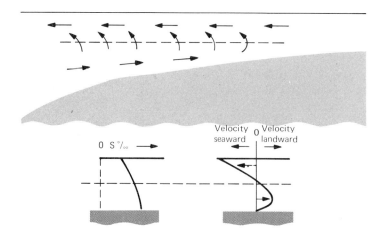

balance exists between the salt transported inward at the bottom and the salt swept seaward by the river. The salt balance tends to be maintained, because a change in the flow pattern is countered by a change in the salinity distribution in the estuary.

Sedimentation within a salt-wedge estuary is controlled by the water circulation pattern. Fine sediments (clay) that are dispersed in the river-water tend to form aggregates when introduced into seawater. This process, called *flocculation*, occurs in water whose salinity is as low as 8 parts per thousand. The flocculated aggregates sink rapidly and tend to accumulate in the saline wedge above the river bed. Accumulation of sediment in the upper reaches of the estuary is also increased by fine silt carried upstream within the saline wedge (Fig. 10-29A).

During the seasons of low river flow, the tidal currents may become dominant in an estuary and tidal mixing occurs at all depths. Under these circumstances, the vertical gradient of salinity is very small. The estuary is, therefore, considered *vertically homogeneous*. However, a horizontal salinity gradient exists, and the salinity increases toward the sea. The pattern of estuarine circulation shown in Fig. 10-29A almost disappears, and accumulation of suspended sediment within the estuary is retarded.

These two conditions (i.e., river-dominated and tide-dominated circulation) simply represent the ends of a continuous spectrum of possible circulation patterns. The ratio of tidal amplitude to river flow determines the pattern that will exist. In most estuaries, the circulation is mixed, falling somewhere between the salt-wedge and the vertically homogeneous cases (Fig. 10-29B). Where river flow changes significantly between seasons, the estuarine circulation changes accordingly. As a result, a given estuary may be closer to the salt-wedge type during peak river runoff and may be well-mixed during low runoff.

A knowledge of circulation is important in the design of waste discharge systems for municipalities and industries located on coastal plain estuaries. An effluent discharged near the bottom of a salt-wedge estuary would migrate upstream and so may not be dispersed quickly by mixing processes.

FJORDS. Fjords are inlets in the coastal reaches of formerly glaciated valleys. They occur at latitudes above 38 degrees, usually on the western side of continents or large islands where Pleistocene glaciers carved their way to the sea. We find them, therefore, in Greenland, Norway, western Siberia, Scotland, British Columbia, Chile, and New Zealand.

Fjords are characterized by steep walls, a U-shaped cross section, and a sill built of glacial drift deposited near the seaward entrance (Fig. 10-30). The almost vertical walls of many fjords are some of the few places in the world where beaches do not exist (Fig. 10-31). Because the sill and the narrow mouth restrict the entrance, the circulation of water within a fjord is often unique, being controlled by its geometry.

FIGURE 10-30
A steep narrow fjord type estuary carved by glaciers and subsequently
drowned during the Holocene marine transgression. The peaks rise for over
2500 m while the maximum depth in this fjord exceeds 350 m.

FIGURE 10-31
The walls of some fjords are vertical. Water depth at the place
where this photo is taken exceeds
350 m.

An example of the circulation within a relatively deep-silled fjord is shown in Fig. 10-32. Most fjords exist in temperate and polar regions where the runoff of fresh water is large. We find, consequently, the estuarine features shown in Fig. 10-32A. Biological productivity in mixed estuaries of this sort is generally kept high because mixing and runoff add nutrients to the surface water layers.

FIGURE 10-32

(A) Circulation and water characteristics in Puget Sound, Washington, a basin characterized by excess precipitation and strong mixing. (B) Idealized sketch of the Black Sea, a stagnant basin. The surface layer extends to depths of approximately 200 m.

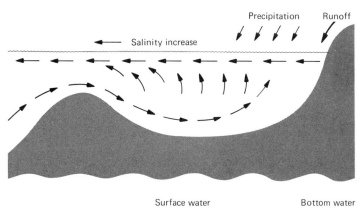

	Surface water	Bottom water
Temp.	5°–15° (winter-summer)	8°–9°C
Salinity	0–32 °/₀₀	32 °/₀₀
A Nutrients	relatively high	relatively high

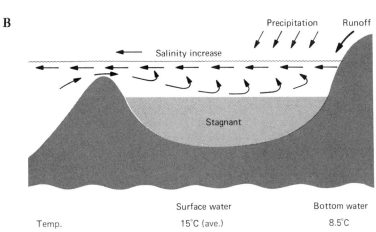

	Surface water	Bottom water
Temp.	15°C (ave.)	8.5°C
Salinity	18 °/₀₀	22 °/₀₀
Oxygen	6 ml/l	0ml/l
Nutrients	moderate	extremely high

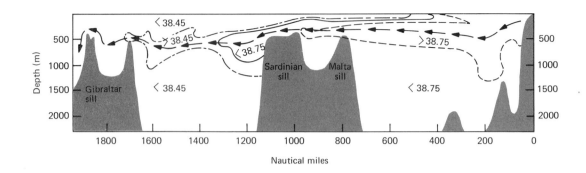

	Surface water characteristics	Bottom water characteristics
Temperature	15° C	13°C
Salinity	39 ‰	38%
Nutrients	relatively low	relatively low

FIGURE 10-33
Circulation and approximate water characteristics in the Mediterranean Sea, a large basin characterized by net evaporation. The longitudinal section of salinity represents the summer condition. The arrows indicate the flow of subsurface water.

Many fjords have extremely shallow sills and great depth (300 to 400 m). These basins tend to become stagnant, because tidal mixing is not sufficient to recirculate deep-water layers. Bottom water becomes *anoxic* when the rate of consumption of oxygen during breakdown of organic detritus settling from the surface layers exceeds the rate of oxygen replenishment. Anoxic water, although rich in the nutrients produced in the decomposition process, is devoid of organisms other than anaerobic bacteria.

The dense bottom water in an anoxic basin can be replaced with oxygenated seawater. Occasionally, when either the runoff of fresh water becomes very low or upwelling brings uncommonly dense water above the sill depth, the body of anoxic water is displaced and kills many marine organisms as it circulates outward. After such an episode of overturning, the process of oxygen depletion in the basin repeats itself. Although it is not a fjord, the Black Sea provides an illustration of this phenomenon.

BAR-BUILT ESTUARIES. On broad, gently sloping continental shelves, sand carried by wave action accumulates in offshore bars aligned parallel to the coast. These bars become barrier islands as they are built upward and shoreward by storm waves that move sand from offshore areas and from sediment sources upcoast. The barrier islands enclose coastal water and form shallow estuaries.

Bar-built estuaries are found on lowland coasts throughout the world. The eastern and gulf coasts of the United States have many good examples. There are several types of circulation and mixing processes

188

within these estuaries. The lower valley of the river that discharges into such an estuary exhibits the characteristics of a salt-wedge estuary. The main part of the bar-built estuary is shallow, with restricted entrances between the barrier islands. As a result, the influence of tides (except within the entrance channels) is minimized. Wind, on the other hand, is an important source of energy for mixing and circulation.

In dry climates where river flow fluctuates dramatically with rainfall, the salinity within the estuary can vary from essentially fresh during high runoff to highly saline (greater than 40 parts per thousand) when net evaporation is excessive.

ESTUARIES PRODUCED BY TECTONIC PROCESSES. In some coastal areas, earth movement by faulting, folding, or local subsidence produces marine basins that receive river discharges. The entrance to these tectonically produced estuaries is often restricted (as in San Francisco Bay). Hence, the circulation and mixing processes can be dominated by either river or tidal flows. In areas having an excess supply of fresh water, the circulation can follow the examples shown in Fig. 10-32.

In regions characterized by net evaporation, the circulation might be reversed, as illustrated in Fig. 10-33. Here, seawater enters the basin on the surface, becomes more dense because of evaporation, sinks, and flows out along the bottom. Biological productivity within an estuary of this sort is relatively low, because the water comes from oceanic surface water already depleted in nutrients. Some nutrients are introduced into the estuary by runoff but not in sufficient quantities to promote phytoplankton growth. Estuaries and larger restricted basins that exhibit a surface inflow of water are found in arid regions. The upper Gulf of California, the Red Sea, and the Mediterranean Sea are examples.

READING LIST

BEACH EROSION BOARD OFFICE OF THE CHIEF OF ENGINEERS, *Shore Protection and Planning.* Tech. Rept 4. Washington, D.C.: Dept of the Army Corps of Engineers, Government Printing Office, 1961. 242p.

BIRD, E.C.F., *Coasts.* Canberra: Australian National University Press, 1968. 246p.

KING, C.A.M., *Beaches and Coasts.* London: Edward Arnold (Publishers) Ltd., 1960. 403p.

LAUFF, G.H., ed., *Estuaries,* American Association for the Advancement of Science Publication 83. Washington, D.C., 1967. 757p.

ONE
TWO
THREE
FOUR
FIVE
SIX
SEVEN
EIGHT
NINE
TEN
ELEVEN
TWELVE
THIRTEEN
FOURTEEN
FIFTEEN

classification and description of marine organisms

Before discussing the principles of biological oceanography, we shall describe the kinds of life found in the sea. To maintain a systematic treatment of the numerous different forms of life, we assume a classification of life forms that is based on a generalized nature of groups of organisms rather than on individual species.

The organisms living in the sea have been studied for centuries. Early observers recognized gross differences between plants and animals, so all life was classed in one or the other of these kingdoms. This system faltered upon the recognition that certain planktonic organisms had the animal characteristic of motility and the plant characteristic of green photosynthetic pigments. The study of bacteria added further confusion, because some are pigmented and others are not, some photosynthesize and others do not. Indeed, certain viruses may behave like crystalline materials rather than like living microorganisms. Clearly, a different means of classifying such deviant forms became necessary.

Studies in cellular biology have provided a basis for one scheme of classification. The organization of the nuclear parts of the cell suggests the separation of *prokaryotes*, which are nonnucleated bacteria and blue-green algae, from the *eukaryotes*, or all forms of life having cells containing definite membrane-bound nuclei.

190

TABLE 11-1

CLASSIFICATION IN THE PLANT AND ANIMAL KINGDOMS

Phylum (or Division)
Class
Order
Family
Genus (Genera)
Species

Through molecular biology, scientists have learned more about the transmission of genetic information and have defined another kingdom—the protistans, which include protozoans, nucleate algae, and fungi like mushrooms, and yeasts. However, for our purposes in this text, it is sufficient to retain the earlier notion of just the plant and animal kingdoms.

The taxonomic subdivisions of the plant and animal kingdoms used to describe the forms found in the sea are shown in Table 11-1.

This system of identifying flora and fauna is called the Linnean system of nomenclature, after the Swedish botanist Linnaeus, who developed it in the eighteenth century. The headings are listed with the most general grouping first (phylum) and the degree of specialization increasing downward. The phylum represents a broad group of organisms having common features, whereas the species represent individual plants or animals. Such a system organizes the great variety of organisms found in the ocean, so that their relationships and evolutionary patterns can be recognized and studied. The systemization of life forms follows the principle that the

FIGURE 11-1
Some common forms of marine algae: (A) *Fucus*, a brown alga; (B) *Ulva*, a green alga; (C) *Sargassum*, a brown alga; (D) *Polysiphonia*, a red alga. (H.U. Sverdrup, Martin W. Johnson, and Richard M. Fleming, *The Oceans: Their Physics, Chemistry, and General Biology*, © 1942. Reprinted by permission of Prentice-Hall, Inc., Englewood Cliffs, N.J.)

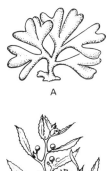

A

B

C

D

ontogeny of an organism recapitulates its phylogeny; that is, the stages of development of an organism represent stages in the evolution of the ancestors of that organism. This principle provides a functional basis for the system of classification that we use today.

11.1 PLANTS IN THE OCEAN

Nowadays, botanists divide the plant kingdom into two subkingdoms, based on method of reproduction: Thallophyta (algae, fungi, bacteria) and Embryophyta (all others). The former is important in the study of oceanography, because all its members are completely dependent on a water environment for reproduction. However, some of the higher plants do live near the sea and can affect coastal—particularly estaurine—environments.

SUBKINGDOM THALLOPHYTA

By far the most important plants in the ocean are the algae (Plate I and Fig. 11-1). These organisms have a very primitive structure and contain no roots, flowers, stems, or leaves. Marine algae inhabit all portions of the world ocean that receive direct solar energy. For instance, algal forms abound on rocks high in the intertidal zone, and as microscopic plankton they drift throughout the world ocean. Their role in the economy of the sea is critical and will be discussed in detail in Chapters 12 and 13.

The primary classification of algae is usually based on the color of the organism, although the color seldom indicates the family or genus to which the plant belongs. These more specific classifications are determined by life history and, to some extent, by anatomical structure and type of food reserves.

Blue-green algae (Phylum Cyanophyta) are microscopic in size and occur in large concentrations both in the marine and freshwater environments. They are more common in fresh water where they form large mats and surface scum having a bad odor. The Red Sea derives its name from a planktonic (drifting) form of this type of alga that is reddish in color. In this case, the organism is classified as a blue-green alga on the basis of its anatomical structure. Blue-green algae reproduce by division of single cells into two smaller individuals. After growth, the *smaller* individuals continue to divide. The cells of the blue-green algae are not nucleated, thus suggesting that they are extremely primitive plants.

Green algae (Phylum Chlorophyta) are sessile (attached) plants that inhabit rocky shores. Examples are sea lettuce (*Ulva*), which occurs in temperate regions, and *Halimeda*, a calcium-carbonate secreting plant growing in abundance in tropical regions and found around atolls. These algae reproduce by spores that are simple in structure and may, or may

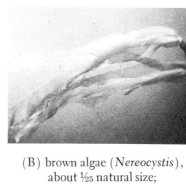

(B) brown algae (*Nereocystis*),
about ⅟₂₅ natural size;

(A) green algae (*Ulva*),
about half natural size;

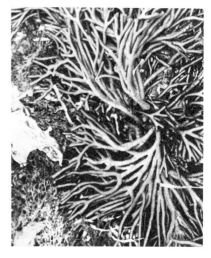

(D) red algae,
about ¼ natural size;

(C) red algae,
about half natural size;

(E) encrusting algae,
about twice natural size.

PLATE I
Characteristic forms of attached marine algae.
(A, C, D, E courtesy of Bernard Nist; B courtesy of Peter B. Taylor).

not, have the power of spontaneous movement. Spores produced at different seasons of the year eventually become free from the parent plant. Provided that they find an attachment place within a reasonable amount of time, they then develop into new plants.

Brown algae (Phylum Phaeophyta) are predominantly marine and have forms of quite diverse size and structure. One example is kelp, a giant seaweed attached to a rocky bottom. It often grows over 80 m long. This type of algae grows in beds or "forests" in offshore water where it provides shelter for a characteristic assemblage of fish and invertebrates and, to an extent, protects the coastline from waves. *Sargassum*, or gulf-weed, is also a brown algae but is much smaller than kelp. It grows in the Caribbean Sea. During storms, it breaks off and is swept into the center of the North Atlantic current gyre, giving the Sargasso Sea its name. The loose masses and tangles of floating seaweed segments are only a few centimeters long. They float freely for a time and then sink when their floats are lost. Other forms of brown algae grow attached to rocks in the intertidal zone. One type is shown in Fig. 11-2.

Red algae (Phylum Rhodophyta) are generally smaller than brown algae. They are sedentary forms and may look like crumpled red paper. Colors range from rose red to deep purple-red or almost black. A variety called *Lithothamnion* secretes calcium carbonate and is a principle organism in cementing carbonate reefs. Another form is a source of agar. Others are used for extracting minerals and for fertilizers. These algae are distributed widely in intertidal and subtidal regions of the world ocean.

Diatoms (Class Bacillariophyceae) are the most abundant of the phytoplankton in the world ocean (Plate II). They reproduce by cell division; and in some cases, one-half of a diatom population reproduces every 24 hours. Diatoms are very small single-celled plants. They are unique because they are enclosed in a shell (or *frustule*) of silica. Many

FIGURE 11-2

The gross structure of the brown algae *Nereocystis*.

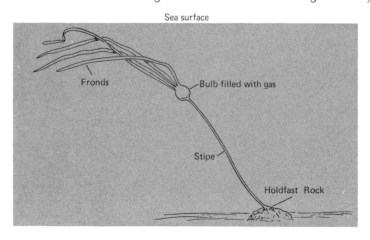

Sea surface

Fronds

Bulb filled with gas

Stipe

Holdfast Rock

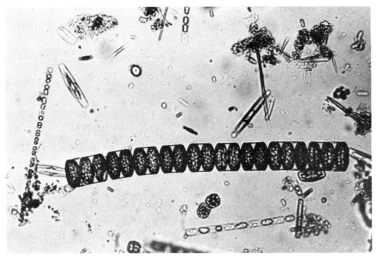

(A) various diatoms, magnification
400 ×;

(B) highly magnified view of a
diatom frustule, magnification
8000 ×;

(C) highly magnified view of a
coccolithophore and separate
plates, magnification 4600 ×;

(D) the Dinoflagellate *Gonyaulax*
responsible for *Red Tides*,
magnification 250 ×.

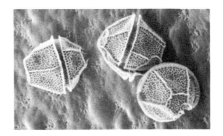

PLATE II
Characteristic forms of planktonic marine algae. (Photograph A courtesy of
Bernard Nist; B, C courtesy of U.S. Department of Interior, Geological
Survey EROS Program; D courtesy of Kent Cambridge Scientific).

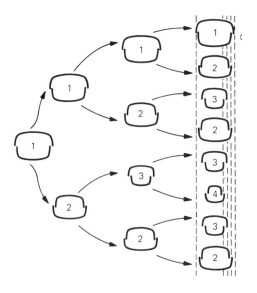

Figure 11-3
Reproduction in diatoms. Dim-
inution of size resulting from cell
division in three generations.
(After Sverdrup, Johnson, & Fleming, 1942)

of these plants have the form of a pillbox or a cigar and are arranged either singly or connected in chains. Generally, the diatom reproduces in such a way that the pillbox separates, and two pillboxes result (Fig. 11-3). Because this form of reproduction leads to increasingly smaller individuals, it cannot continue indefinitely. After a number of cell divisions, an *auxospore* is formed within which a cell separated from its tiny shell may grow and form a full-sized frustule. Resting spores can also be formed; these spores allow the cell to survive unfavorable environmental conditions. Diatoms survive only by remaining in the upper sunlit regions of the world ocean. These plants are well adapted for spending their lives in the surface waters (see Chapter 13). However, benthic (bottom-dwelling) forms do exist in shallow water, and other forms grow on plants and animals. In many regions of the ocean, the silica shells, or frustules, of diatoms have accumulated in great numbers in marine sediments.

Dinoflagellates (Class Dinophyceae) are single-celled organisms that, as a group, possesses both plant and animal characteristics (Plate II). Many of these organisms photosynthesize as plants do, but others move and ingest food as animals do. These organisms have a shell of cellulose. The pigmented forms are considered phytoplankton. Although they possess two hairlike appendages (*flagella*) for locomotion, they are only weakly mobile.

Coccolithophores are among the smallest phytoplankton in the open ocean (Plate II). They possess two flagella and are also weakly motile. The soft parts of these single-celled organisms are covered with small calcite plates called *coccoliths* or rabdoliths. These plates separate

196

from the cell when the organism dies and accumulate on the sea floor as a biological component of marine sediment.

SUBKINGDOM EMBRYOPHYTA

The higher plants, which have true roots and produce seeds, live in the upper subtidal regions (from low tide to a depth of 7 m). These plants actually evolved on land and invaded the shallow-water of the sea. Examples of higher plants living in the world ocean are mangrove trees, *Zostera* or eel grass, and *Phyllospadix* or surf grass. These plants inhabit the shallow water of many tidal flats, estuaries, and bays (Fig. 11-4).

11.2 ANIMALS IN THE OCEAN

There is a great variety of animal life in the ocean. Representatives of all major animal groups, except amphibians, are found in the world ocean. In fact, many animal groups are exclusively marine. This variety of animal life is much greater than on land or in fresh water. It depends, however, upon latitude; there is more variety in the tropical ocean than in the polar seas. But most of the animal forms are small; the average marine organism is smaller than a mosquito. In the following discussion, only those forms particularly relevant to oceanography are included.

PHYLUM PROTOZOA

Protozoa are small organisms that perform all their life processes in a single cell. Planktonic (drifting) and benthic (bottom-dwelling) forms inhabit all parts of the world ocean (Plate III). Of particular importance are those protozoa that secrete or build shells, which are called *tests*. These tests show a wide diversity of structure and are composed of calcium carbonate ($CaCO_3$), silica (SiO_2), or detrital material cemented with calcium carbonate. The organisms extend their protoplasm outside

FIGURE 11-4

The eel grass *Zostera* on an intertidal sand flat. The strands are about 1 m long. (Photograph courtesy of Peter B. Taylor).

(A) various Radiolaria tests,
magnification 100 ×;

(B) Foraminifera tests,
magnification 45 ×;

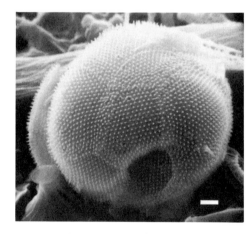

(C) highly magnified view of a
Globigerina, magnification 7000 ×;

PLATE III
Characteristic forms of planktonic protozoa.
(Photographs A, courtesy of Kent Cambridge Scientific; B, courtesy of Y. R.
Nayudu; C, courtesy of U.S. Department of Interior, Geological Survey
EROS Program.)

of their tests in order to capture food. The dinoflagellates and other motile algae are sometimes included in this group of organisms.

Order Foraminifera is a group of protozoa that is both benthic and planktonic. However, the benthic forms are not prevalent in the open ocean areas. The planktonic genus *Globigerina*, for example, abounds in the warmer or subtropical surface waters of the open ocean and can live at higher latitudes. It secretes a calcium carbonate test consisting of one or more spherical chambers. Upon death or reproduction, these chambers sink to the sea floor and become a recognizable portion of the sediments.

Order Radiolaria is an important group of planktonic protozoans that secrete a silica test. These tests are extremely intricate and varied in their form and structure. Radiolaria are distributed throughout the arctic, antarctic, and tropical surface regions of the world ocean and contribute significantly to the sediment at the sea floor.

PHYLUM PORIFERA

The Porifera, or sponges, are clusters of cells having a loose organization. Sponges are attached to the bottom, and they live at all depths. They have few natural enemies, they are filter feeders, and some have a planktonic larvae. The skeleton of a sponge consists of needle-shaped spicules oriented more or less at random throughout the protoplasm. The spicules are made of a variety of substances: calcium carbonate, silica, and a woody material similar to cellulose (Fig. 11-5).

FIGURE 11-5

(A) A colony of sponges, magnification about 4 ×; (B) a sponge spicule approximately 1 mm long. (Photograph A, courtesy of Bernard Nist; B, courtesy of Kent Cambridge Scientific.)

A
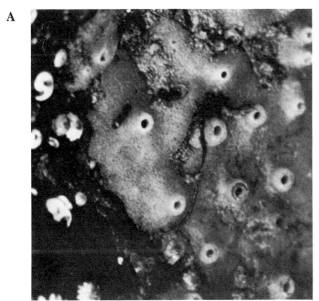

B

(A) small hydroid polyp;

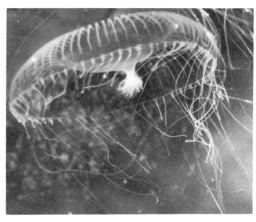

(B) medusa;

(C) sea anemones in a tide pool. One
exposed specimen has its tentacles retracted;

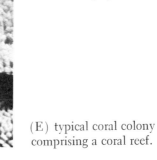

(D) individual coral polyps

(E) typical coral colony
comprising a coral reef.

PLATE IV
Characteristic coelenterates about natural size.
(Photographs A, B, D courtesy of Bernard Nist; E courtesy of Peter B. Taylor).

PHYLUM COELENTERATA (OR CNIDARIA)

The Coelenterata (Plate IV) are fairly simple organisms that consist of a gut lined with protoplasm with an opening at one end ringed with tentacles. Coelenterates may be either stationary (hydroid) or planktonic (medusa). The stationary hydroid form is a common tide pool inhabitant. Jellyfish have the medusa form. This phylum contains three major classes. The *Hydrozoa* are the hydroid polyps and some medusae, little jellyfish about 2 centimeters in diameter. Class *Scyphozoa* contains the true jellyfish, free-swimming medusae that inhabit all parts of the world ocean. They sometimes attain a diameter of 2 m with tentacles extending 30 m below the umbrella. The *Anthozoa* is the largest class; it includes the sea anemones and corals. All members of the Coelenterata are carnivores and possess stinging cells called *nematocysts* in their tentacles. These cells look like a small bladder filled with a coiled, hollow needle, tipped with barbs. When a trigger at one end of the cell is touched, the needle fills with water and distends itself from the cell, thus injecting anything nearby with a poisonous secretion. In tropical seas and along the eastern coast of the United States, a jellyfish known as *Physalia* (Portuguese man-of-war) is a nuisance to swimmers, because it can inflict severe pain to those who touch its tentacles. In the tropics, there is also a jellyfish called a sea wasp whose sting is often fatal.

A

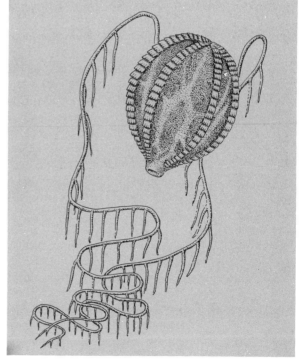

FIGURE 11-6
Photograph (A) and line drawing (B) of a comb jelly. The body is approximately 1.5 cm long. (Photograph courtesy of Bernard Nist).

B

PHYLUM CTENOPHORA

Ctenophora are marine organisms that attain diameters of only a few centimeters. Most of these organisms are planktonic; and all of them are bioluminescent—that is, they are able to emit light. This luminescence is often seen at night in the wake of a ship. Ctenophora have eight rows of comblike appendages used as paddles while swimming. They also have two long tentacles that contain sticky cells used in capturing prey. Some members of this group are called comb jellies, sea walnuts, or sea gooseberries in accord with their appearance (Fig. 11-6). A very beautiful member of the Ctenophora, Venus's-girdle, is ribbon-shaped and appears iridescent in the proper light.

PHYLUM PLATYHELMINTHES

This division is made up of the flatworms. Some of these organisms are leaflike in shape and are often attractively colored (Plate V). They live on the bottom of the ocean among rocks or crevices and burrow into the mud. Movement is by means of hairs (*cilia*) that cover the body of the organism.

PHYLUM MOLLUSCA

The organisms in this division have soft bodies; some also have a rasping tongue called a *radula*. Many molluscs have a calcium carbonate shell. This shell is secreted by a layer of tissue called the *mantle*. Organisms in the Mollusca are so diverse that they are separated into several classes (Plate VI).

Class Amphineura includes the chitons that inhabit tide pools and rocky coastlines. These organisms have soft bodies with eight separate plates of shell material over their backs. They feed by scraping algae off rocks.

Class Gastropoda includes the snails and slugs found on land, in fresh water, and in the sea. Gastropods are generally benthic as adults and, by using a large "foot," can creep along the sea floor. In some species, the foot has evolved into a swimming organ, and these gastropods are wholly planktonic. A notable example is the *pteropod* that is frequently found in tropical oceanic waters. The calcareous portions of pteropod shells contribute to marine sediments in areas where these organisms are abundant.

Class Pelecypoda (or Lamellibranchia) is made up of clams, oysters, and mussels. These are bivalve organisms; that is, their calcium carbonate shells are fastened together along a hinge line. Pelecypods are filter feeders. As adults, they are benthic, living attached to rocks and pilings or burrowing beneath rocks, pilings, and sediments on the sea bottom.

Class Cephalopoda includes the octopus, squid, and chambered nautilus. These organisms have a parrot-like chewing beak, tentacles with suckers for catching prey, and a water-jet propulsion apparatus. The

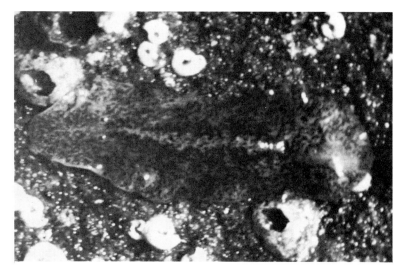

(A) a flatworm well camouflaged
on a tide-pool rock,
about four times
natural size;

(B) a polychaete worm with
tentacles exposed, about
twice natural size.

PLATE V
Characteristic marine worms.
(Photographs courtesy of Bernard Nist).

(A) chiton,
natural size;

(B) limpet,
about twice natural size;

(C) pecten,
natural size;

(D) nudibranch,
about twice natural size;

(E) a benthic octopod
photographed at a depth
of 4000 m,
about 1/20 natural size;

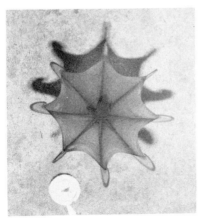

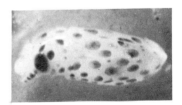

(F) juvenile squid,
about natural size.

PLATE VI
Characteristic forms of molluscs
(Photographs A, B, C, D, F courtesy of Bernard Nist; E courtesy of
U.S. Naval Research Laboratory).

squids have remarkably good eyes and are very fast swimmers. They swim tentacle-first and when startled will propel themselves backwards rapidly by squirting water from an opening in their bodies. Cephalopods can attain a length of 16 m.

PHYLUM ANNELIDA

This phylum includes the various kinds of segmented worms (Plate V). The largest class is *Polychaeta*, whose members resemble the common earthworm, although many have bristly fans or tentacles and build tubes that stand up, giving the appearance of a flower. These organisms are mainly benthic. The planktonic forms have rows of paddle-like appendages down the sides for swimming.

PHYLUM ARTHROPODA

Arthropods are joint-legged organisms. On land, examples are the insects, spiders, and scorpions. In the oceans, the various arthropods are well developed, diverse, and extremely numerous (Plate VII). Taxonomically, these marine forms belong to the Class *Crustacea*, which has a number of subdivisions. Only a few of the more important oceanic groups of crustaceans are discussed here.

Subclass Cirripedia includes the barnacles found in all benthic environments. Barnacles live inside calcareous enclosures that are attached to rocks, pilings, or on floating materials, such as ships, whales, logs, and net-floats. Some forms build their own floats and drift like plankton. These organisms resemble crustaceans only during their larval stages when they closely resemble copepod larvae. Barnacle larvae must settle on a suitable surface in order to complete their adult growth, hence they are dependent on the presence of solid objects.

Subclass Copepoda contains some very small organisms, about 2 millimeters in length, which make up the bulk of the zooplankton in the world ocean. They are, therefore, probably the most numerous of all multicelled animals. Copepods are important because, as grazers of phytoplankton, they convert the tiny plants of the ocean into animal matter that is utilized subsequently by still larger animals. They feed on phytoplankton by trapping them in numerous bristles on their larger appendages. The bristles form a basket-like apparatus near the mouth that filters water and collects phytoplankton as the copepod swims.

Order Euphausiacea is made up of shrimp-like organisms that attain a length between 1 and 5 cm and have luminescent spots along their sides. These organisms are abundant as plankton and are also found at, or near, the sea floor. Euphausids, frequently called krill, are the principle food for filtering-type whales, such as the baleen whale.

Order Decapoda includes several subgroups. These are lobsters, crabs, hermit crabs, shrimps, and prawns. Most of the adult decapods are benthic and provide a source of food for humans.

(A) Acorn barnacle dormant;

(B) Goose barnacle with feet in feeding position;

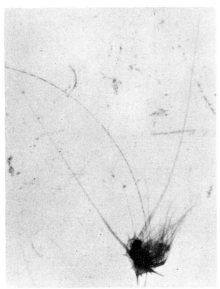

(D) crab;

(C) the barnacle larvae
(body diameter 2 mm);

(F) the copepod *Calanus* (length
approximately 2mm);

(G) *Euphausia pacifica*
(length approximately 3 cm).

(E) shrimp;

PLATE VII
Characteristic forms of arthropods about natural size.
(Photographs A, B, D, E courtesy of Bernard Nist;
C courtesy of Dora P. Henry).

PHYLUM ECHINODERMATA

Echinoderms are spiny-skinned organisms that are exclusively marine and are generally bottom dwellers (Plate VIII). Commonly known examples are starfish (or sea star), brittle star, sea urchin, and sand dollar. Some of these organisms are filter feeders (sea cucumbers), but most eat algae (sea urchins) or clams (starfish). They all have radial symmetry, usually with five parts in the shape of a star.

PHYLUM CHAETOGNATHA

This phylum contains the arrow worms, which are exclusively marine planktonic organisms (Fig. 11-7). They are voracious eaters and feed largely on copepods. These organisms have been used in tracing water masses, because the arrow worms in one water mass are distinctively different from those in another.

PHYLUM CHORDATA

Chordates are animals that have, at some time in their development, gill slits and a cartilaginous skeletal rod known as a *notochord* (Plate IX). Although there are several subphyla, only two are significant for our purposes. The lower chordates in Subphylum Urochordata or Tunicata, have no backbone and are exclusively marine organisms (Plate IX A).

FIGURE 11-7
(A) Photograph and (B) sketch of the arrow worm
Sagitta (body length approximately 2 cm.)

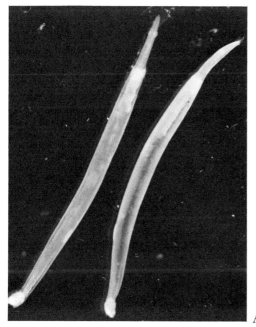
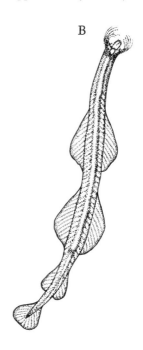

A

(A) and (B) starfish,
about 1/5 natural size;

(C) sea urchin,
about 1/3 natural size;

(D) sea cucumber,
about 1/2 natural size.

PLATE VIII
Characteristic forms of echinoderms
(Photographs courtesy of Bernard Nist).

(A) tunicate,
about 1/2 natural size;

(B) green sea turtle,
about 1/10 natural size;

(C) porpoise and killer whale;

(E) part of a whale baleen,
about 1/10 natural
size.

(D) jaw and teeth of a killer whale,
a man's hand provides the scale;

PLATE IX
Characteristic forms of chordates.
(Photographs B, C courtesy of Bernard Nist; E courtesy of F. Raco,
Brooks Institute).

Some are benthic filter feeders (sea squirts or tunicates), and others are planktonic (*Oikopleura*). *Oikopleura* builds a fragile gelatinous body and moves through the water by whipping its tail. As it moves, it filters water through a screen within its body. When this screen becomes full of captured material, it eats the filter, leaves its gelatinous body, and builds a new one.

The vertebrate chordates are members of Subphylum Vertebrata, which is made up of several important and familiar subgroups or classes.

Class Chondrichthyes consists of primitive fishes (sharks and rays) that have a skeleton composed of cartilage. These organisms grow very large. The whale shark is the largest of all fishes and attains a length of over 20 meters.

Class Osteichthyes contains true fishes that have a bony skeleton. They are generally streamlined, carnivorous, and live throughout the marine environment. Fishes show exceptional degrees of development, depending on their habitat in the ocean. In fact, they serve as an excellent example of environmental adaptation and are discussed in Chapter 13.

Class Reptilia includes some air-breathing snakes that are found only in surface waters. Sea snakes 2 m long inhabit tropical and semitropical waters and are very poisonous. Sea turtles often attain weights exceeding 400 kilograms (Plate IX B).

Class Aves includes a number of birds that are dependent upon the sea for food—for example, penguins, pelicans, and gulls. Some birds, such as the albatross, petrel, and auk, return to land only for nesting.

Class Mammalia includes a number of mammals that are also closely associated with the ocean. Polar bears, sea otters, seals, walruses, sea lions, and sea cows are more adapted for life in a marine environment than life on land. Whales and porpoise (Plate IX C) are exclusively marine mammals. Baleen (whalebone) whales have a series of long, frayed plates hanging in their mouths (Plate IX E). The plates filter large volumes of water in order to capture plankton. The blue whale attains a length of over 30 meters and a weight of about 100,000 kilograms. The killer whale is also a member of this subgroup. Some mammals in this category have teeth (Plate IX D). The sperm whale has teeth only in its lower jaw. Dolphins and porpoises have teeth in both jaws.

READING LIST

BUCHSBAUM, R., *Animals without Backbones* (rev. ed.)   Chicago: University of Chicago Press, 1948.   405p.

DAWSON, E.Y., *How to Know the Seaweeds.*   Dubuque, Iowa: W.C. Brown Co., 1956.   197p.

HERALD, E.S., *Living Fishes of the World.*   Garden City, N.Y.: Doubleday & Company, Inc., 1961.   303p.

NICOL, A.C., *The Biology of Marine Animals.*   New York: Interscience, 1960. 707p.

ZOBELL, C.E., *Marine Microbiology.*   Waltham, Mass.: Chronica Botanica Co., 1946.   240p.

ONE
TWO
THREE
FOUR
FIVE
SIX
SEVEN
EIGHT
NINE
TEN
ELEVEN
TWELVE
THIRTEEN
FOURTEEN
FIFTEEN

introduction to biological oceanography

12.1 DEVELOPMENT OF LIFE IN THE SEA

Life probably began in the sea. The diversity and the antiquity of marine organisms support this conclusion. Major subdivisions of both plant and animal kingdoms have many representatives in the world ocean. In fact, five of the animal phyla are exclusively marine, and only 6 per cent of the invertebrates are exclusively nonmarine.

Fossil evidence indicates that blue-green algae developed photosynthesis and produced oxygen between 1.8 and 2.5 billion years ago. As a result, the earth changed from a reducing environment to an oxidizing one. Prior to that time, primitive, nonphotosynthesizing bacteria prevailed. After the change to the oxidizing environment, green plants and animal prototypes evolved. Marine invertebrates thrived 400 to 500 million years ago. Some forms, such as the brachiopod (a marine mollusk-like organism), have remained unchanged to the present time, whereas many other

FIGURE 12-1

Geologic Time Chart. Ascending lines indicate the range in time of most of the chief groups of marine animals and plants. If the line ends in a crossbar, this denotes the time of extinction; if it ends in a dart, the group is still living. (After Dunbar, *Historical Geology*, 1949, John Wiley & Sons)

212

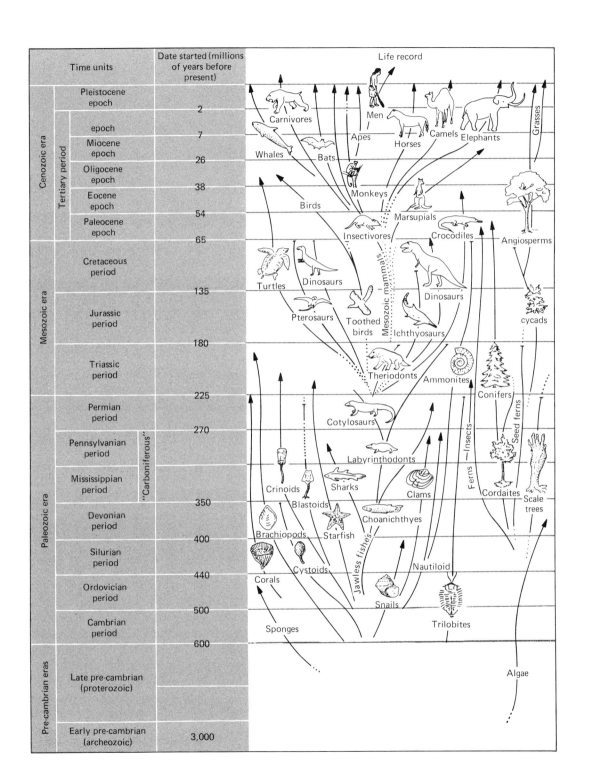

	Time units		Date started (millions of years before present)	Life record
Cenozoic era	Tertiary period	Pleistocene epoch	2	
		epoch	7	
		Miocene epoch	26	
		Oligocene epoch	38	
		Eocene epoch	54	
		Paleocene epoch	65	
Mesozoic era	Cretaceous period		135	
	Jurassic period		180	
	Triassic period		225	
Paleozoic era	Permian period		270	
	Pennsylvanian period	"Carboniferous"		
	Mississippian period		350	
	Devonian period		400	
	Silurian period		440	
	Ordovician period		500	
	Cambrian period		600	
Pre-cambrian eras	Late pre-cambrian (proterozoic)			
	Early pre-cambrian (archeozoic)		3,000	

organisms have experienced continuous evolution (Fig. 12-1). Fish appeared in the sea during the early Paleozoic era, and amphibians invaded the land in the late Paleozoic (300 million years ago). Mammals evolved on land about 225 million years ago; as a group, they gained ascendency during the Cenozoic era (from 65 million years ago to the present). Some mammals, like the whale and the porpoise, have evolved back into the sea.

The influence of the environment upon the development of life in the sea is reflected in the chemistry of body fluids in many marine animals. The concentrations of major ions in seawater are similar to those in the body fluids of many marine invertebrates. This observation supports the idea that seawater developed its salinity early in its history (see Chapter 4) and that the composition of seawater is fundamental to the metabolic processes of primitive organisms.

12.2   LIFE REQUIREMENTS OF MARINE ORGANISMS

The requirements for life in marine organisms are similar to those of terrestrial forms; however, significant differences do exist. Representing an evolutionary response to the environment, these differences are noticeable in the morphology and physiology of various groups of organisms. The important life requirements and a short discussion of how organisms respond to these needs are listed below.

WATER

Organisms in the sea live in close association with their supporting medium. Seawater surrounds or permeates the body cavities and internal systems of the invertebrates, so it is not necessary for them to possess mechanisms to safeguard against *desiccation* (excessive loss of water) unless they live in the intertidal region. In most of the vertebrates, regulation of water intake is required, because their body fluids are less saline than seawater. Since there is a tendency for solutions of different concentrations to reach a common concentration by the diffusion of water and dissolved substances, water usually leaves the vertebrate tissue by *osmosis*. In the process of osmosis, diffusion takes place through the semipermeable membrane (i.e., cell wall) that separates the two solutions. Water molecules can pass through the membrane, but dissolved substances cannot; only diffusion of water, therefore, can occur. Most marine vertebrates have regulatory mechanisms for taking in or giving off enough water to maintain a constant internal salinity, otherwise they would die by desiccation or flooding. The urinary system of fishes is one example of internal salinity regulation in marine organisms. Another example is found in

crabs, which maintain their internal salinity by depositing excess salts resulting from metabolism in their carapaces, or outer covering. These
animals shed their carapaces as they grow, thus excreting the excess salts.

SUNLIGHT

Sunlight penetrates the sea to just a limited depth; approximately
1 per cent of the light striking the surface reaches a depth of 100 m. As a
result, marine plants must remain near the surface where light is plentiful
if they are to flourish. Marine algae, therefore, have a particular structure
and mode of living: they are planktonic, primitive, microscopic in size, and
exhibit a variety of shapes. However, response to light is a complex phenomenon. Some marine animals respond to light by avoiding it; some are
attracted to it. Other organisms exhibit daily movements suggesting that
they seek an optimum light intensity.

HEAT

The high heat capacity of water imparts a great degree of stability
to the temperature of the sea. Daily and seasonal temperature fluctuations
do not extend deeply into the sea. For example, the seasonal variation of
temperature in the North Atlantic Ocean is minimal below a depth of 200
m (Fig. 12-2). Such temperature stability assures that marine organisms
of the open ocean do not experience temperature variations. Accordingly,
marine organisms do not possess physiological temperature regulating
mechanisms.

FIGURE 12-2

Annual variation in temperature with depth in the North Atlantic.
(Data from Walford and Wicklund, American Geographical Society, 1968)

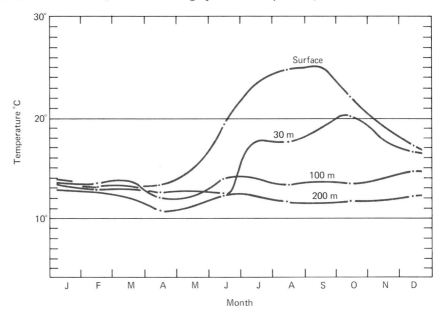

OXYGEN AND CARBON DIOXIDE

Dissolved oxygen and carbon dioxide are abundant throughout the sea, except in enclosed basins where circulation is restricted, and possibly at the oxygen minimum layer. These gases are important to life in the sea. Plants require CO_2 for photosynthesis, whereas both plants and animals consume O_2 during respiration, which is the process that organisms use to oxidize matter and obtain the energy for life. Both photosynthesis and respiration are discussed in detail in Chapter 13. In this section, we only consider how animals and plants obtain O_2 and release CO_2.

Living cells obtain energy by the oxidization of organic molecules that are either absorbed directly or provided by ancillary digestive and circulatory systems. The breakdown of food molecules occurs within the cell, so there must be a mechanism for transferring oxygen from the surrounding environment into the organisms and for removing carbon dioxide, the waste product of metabolism.

In the sea, organisms exchange gases in several ways. Ultimately, the transfer of dissolved substances depends on diffusion of molecules from a region of high chemical concentration to a region of low concentration. In the simplest case, diffusive exchange between the cell and its surroundings occurs directly across the cell wall. This type of exchange occurs in plants and animals having a low degree of organization (i.e., protozoa, sponges, coelenterates, flatworms). Complex marine animals have special organs—gills in fishes and lungs in mammals—for diffusive exchange.

Gills are membranes that contain many blood vessels. When water moves past the surface of the membrane, gases are exchanged between the water on one side of this membrane and the blood vessels on the other side. Oxygen absorbed by the blood is circulated to the cells; carbon dioxide produced within the body cells diffuses into the blood and is carried to the gills where it diffuses into the water. This method of exchange is a three-step process: diffusion-circulation-diffusion. It is used by fishes, crustacea, mollusks, and some worms.

The higher animals in the sea (marine mammals) have lungs for breathing air from the sea surface. Lungs are a complex system of membrane and blood vessels that provide an efficient means of transferring gases from the air into the blood.

FOOD

All forms of life require food substances to form their tissue and to provide a source of energy for their life processes. A variety of food exists in the sea. Marine organisms, therefore use many kinds of food and have several means of acquiring nutrition.

The simplest organisms in the sea, the *autotrophs*, are self-nourishing and produce their own food from H_2O, CO_2, and radiant energy. These organisms are the green plants and certain bacteria.

Heterotrophic organisms cannot synthesize food from inorganic

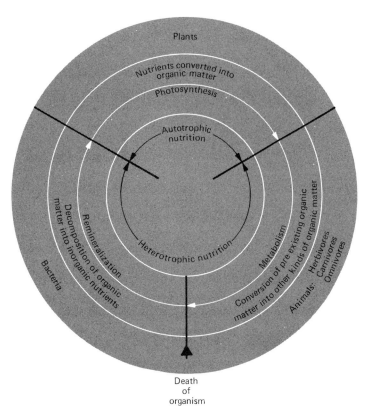

FIGURE 12-3

Comparison of types of nutrition, nutritional processes and
organisms using these processes.

nutrients but must eat other organisms or organic molecules. Hetero-
trophs include all animals, most bacteria, and some plants. Three types
of heterotrophic nutrition are used by animals: holozoic, saprozoic, and
parasitic.

The process of catching, eating, and digesting food particles is
called *holozoic* nutrition. Holozoic organisms have developed sensory,
nervous, and muscular organs to use in capturing food particles. They
have also developed digestive systems to break food particles into useful
nutrient chemicals. In animals, selective diets have evolved; consequently,
animals are grouped on the basis of this selectivity. *Herbivores* are animals
that feed on plants; *carnivores* feed on animals; and *omnivores* feed on
both plants and animals. The relationship between organisms and their
feeding habits is shown in Figure 12-3.

Saprozoic organisms cannot ingest particulate foods but rely on the
direct absorption of organic molecules for nourishment. This process re-

217

quires that these organisms live in an environment containing decomposed animal or plant matter. This form of nutrition is not common among the animal kingdom; it is restricted mainly to the protozoa (single-celled organisms).

Parasitic nutrition occurs when one organism lives on or within another organism and obtains its food at the expense of the host. Almost every animal is the host for one parasite or more. Some parasites cause little or no injury to the host; this relationship is called *commensalism.* Other parasites, however, do great harm by destroying cells, removing nutrients, or producing toxic waste products. Some parasites provide a benefit to the host in one respect and a disadvantage in others. One example is the deep-sea angler fish, which is a sexual parasite (Fig. 12-4). The male fish is permanently attached to the female, so fertilization of eggs is assured but at the expense of nourishment from the female's bloodstream.

SUPPORT

The protoplasm of most animals must be protected from the environment by a covering or a structure that supports the organs and muscle tissue (i.e., a skeleton). Since the density of protoplasm of marine animals is close to that of seawater, the necessity of extensive structural body parts is minimized, and many forms have meager skeletons or none at all. Examples are the jellyfish, squid, and octopus. Even the skeleton of a whale would not be adequate to support its body weight in air.

On the other hand, massive skeletal structures are needed by organisms living in the rigors of the littoral environment where wave action and strong currents would quickly damage animal protoplasm. Conse-

FIGURE 12-4

A female angler fish (*Photocorynus spiniceps*) with a degenerate male fused to her body. The female nourishes the male and he in turn fertilizes her eggs. A luminous fishing organ can also be seen on the front of the female's head. (After Sverdrup, Johnson & Fleming, 1942.)

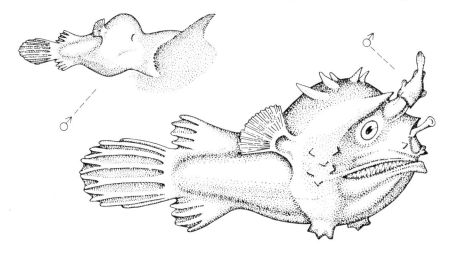

quently, thick external shells, or *exoskeletons,* protect littoral species of clams, snails, and crabs.

Animals that swim need strong muscles attached to a flexible support. Fishes have *endoskeletons* inside their bodies and protective scales (or skin) outside. This arrangement of support and protective structures provides an optimum degree of mobility for these animals.

PROTECTION FROM ENEMIES

Most of the sea is a place without shelter. It is true that there are areas for concealment in the rocks, mud, and seaweeds at the margins of the sea, but most organisms live in open water and therefore are unprotected.

Concealment through protective coloration or transparency helps some of the small invertebrates escape predators. However, organisms such as the copepods, anchovies, and the larvae of most marine organisms cannot outsmart, outswim, or avoid predators, so their survival depends upon their rate of reproduction. Such organisms produce tremendous numbers of offspring, few of which need to survive in order to propagate the species. For example, codfish can produce 5,000,000 eggs at a time, even though only one pair of eggs has to reach maturity in order to replace the parents. If several pairs survived, the eventual production of cod would be overwhelming were it not for the fact that a larger population of predators would be supported. The balance between predator and prey is maintained in such a way that the survival of the species of prey is assured.

There is a premium on simplicity and diversity in the sea. Generally, marine forms are both morphologically and physiologically simpler than their terrestrial counterparts. Furthermore, there is a tendency for primitive marine species to be preserved once they have evolved, probably because the marine environment has remained so uniform during most of its history. Hence, marine life has both great simplicity and great diversity.

12.3 CLASSIFICATION OF MARINE ENVIRONMENTS

Many, but certainly not all, of the individual marine environments recognized by oceanographers are illustrated in Fig. 12-5. We shall consider only the major environments; the others are included for reference. The environments shown in this figure are defined by unique chemical and physical properties and geographic configuration. Since marine organisms respond to these conditions, classification of environments is very meaningful in a biological sense. When individual environments are defined from a biological point of view, they are called *biotic* units. Two such biotic units are recognized in the *photic* (lighted) and aphotic (dark) regions

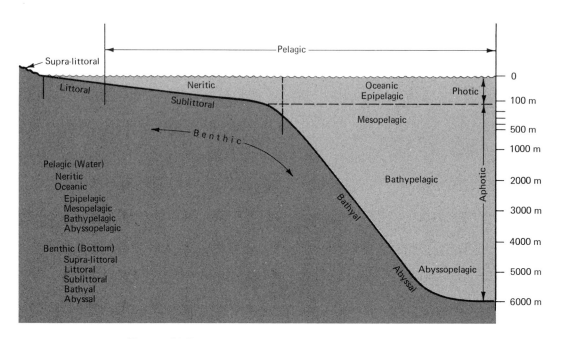

FIGURE 12-5

Classification of marine environments. (After Hedgpeth & Ladd, 1957.)

of the ocean. The boundary line is placed somewhat arbitrarily at 100 m, which is the limit of light penetration in seawater. The photic zone is an important biotic unit, for it is there that plants, the ultimate source of food for all organisms, combine sunlight, CO_2, and nutrients to produce organic matter.

PELAGIC ENVIRONMENT

The term *pelagic* refers to the water of the world ocean. The pelagic environment is divided into provinces both horizontally and vertically. The *neritic province* is the water over the continental shelf, and the *oceanic province* includes all water seaward of the shelf break. The oceanic province is divided vertically into a photic zone and an aphotic zone. The boundary between the two corresponds to the approximate depth of light penetration and to the depth at the outer edge of the continental shelf.

There are basic differences between the environmental conditions in the oceanic and neritic provinces, even though they overlap. In the open ocean, physical conditions—particularly temperature and salinity—do not vary a great deal. Light penetration is also somewhat greater in the oceanic than in the neritic regions (Chapter 5). The neritic province receives a greater continental influence than the oceanic province does. River runoff causes water on the continental shelf to have lower salinity and greater seasonal variation in both temperature and salinity. There is a large supply of nutrients from land; furthermore, nutrients are regenerated on the seabed and welled up from depth in the neritic province. These factors make this area highly productive of plants and animals. So most of the world's fisheries, therefore, are on the continental shelf. The abundance of marine life in the neritic province is as striking as the paucity of life in the subtropical open ocean.

BENTHIC ENVIRONMENT

This type of environment is found at the bottom of the world ocean regardless of depth. It is divided into the *supralittoral, littoral, sublittoral*, and *deep-sea* regions (Fig. 12-5).

The supralittoral environment is that area of the beach above the extreme high water line. This region is rarely submerged; however, it is subject to sea spray and the influence of large storm waves.

The littoral environment is in the intertidal region. This environment is extremely variable. In high latitudes, for instance, temperatures could be subzero in winter (meaning ice) and 40° C during warm summer periods. The salinity in tide pools ranges from zero to greater than 40 parts per thousand, depending on the degree of exposure, evaporation, runoff, and tidal range. There are vigorous water movements in the littoral province. The intertidal zone is continually pounded by waves and swept by currents. Materials at the bottom—rocks, sand, or mud—sometimes change seasonally. Many beaches that are sandy in the summer become rocky during the winter when large waves transport the beach sand to offshore regions.

The sublittoral environment extends from below the tidal zone to the outer edge of the continental shelf. This environment is the transition between the shore and the deep sea. For example, the benthic region of the inner shelf can have daily temperature and salinity changes, but the outer portions vary seasonally. The bottom sediment of the sublittoral includes rock, coral, sand, or mud, depending on conditions of sedimentation (see Chapter 14 for a discussion of marine sediments).

The deep-sea benthic regions are characterized by extensive deposits of fine sediment, uniform conditions of temperature and salinity, no light, and extreme pressures. Seasons do not exist and the values of ambient temperature and salinity are related to the distribution of water masses (Chapter 8).

12.4 MODES OF LIFE IN THE SEA

Contrary to the beliefs of early oceanographers, life is found throughout the marine environment and is as varied as the physical and chemical conditions that prevail. There are three important modes of life in the sea: planktonic, nektonic, and benthic. Marine organisms are classified within these modes according to their swimming ability and habitat.

PLANKTON

Plankton are those organisms that drift passively with the ocean currents. Some organisms in this category can swim weakly, but they spend most or all of their life subject to the motion of the waters that surround them.

Over 15,000 different organisms can be considered as plankton. Members of this group range in size from *nanoplankton*, plants smaller than 60 microns (60/1,000,000 m), to *megaloplankton*, such as large jellyfish whose diameters reach 2 m. Floating plants in the sea are called *phytoplankton*; floating animals are *zooplankton*. Many organisms drift only during their egg or larval stages (*meroplankton*), but others spend their entire life adrift (*holoplankton*).

The existence of a meroplankton phase is an interesting and important aspect of marine life. The majority of marine animals, including fishes, begin as plankton. This mode of life affords them a mechanism of dispersal and survival in a changing environment. Since meroplankton are seasonal, the occurrence of a particular species depends upon its spawning habits. The planktonic larval period can last only a few hours (as in certain worms) or four or five months (as in sand crabs). The larvae of many species spend this time drifting to a place where they can attach themselves for their adult life. Most of the meroplanktonic larvae do not survive; they become food for other marine organisms.

NEKTON

This category includes adult squids, fishes, marine mammals (i.e., whales, seals, porpoise), and larger crustaceans (shrimp-like organisms). The difference between plankton and nekton is indistinct in certain cases. For example, the eggs and larvae of fish are planktonic, but fish become nekton at an early stage in their life.

The term *nekton* implies complete freedom of movement; indeed, nekton are found in all parts of the world ocean. However, certain organisms appear to be restricted to particular geographical locations and depths. There are, consequently, geographical limitations in the distribution of various fisheries throughout the world. For example, the commercially important cod inhabits northern waters of both the Atlantic and Pacific Oceans. These fish migrate freely in these waters; however, they live near the bottom and are rarely caught in depths greater than 350 m. Hence, their habitat is limited in both a geographical and vertical sense.

The boundary that limits the distribution of a species need not be a physical barrier. It can be an environmental one, such as changes or levels in temperature, salinity, oxygen supply, nutrient supply, character of the substrate, or any of many physical and chemical properties. Cod swim off the northeastern coast of the United States only as far south as there is water cooler than about 30° C, the temperature of coagulation of cod albumen. The thermal boundary for cod coincides with the edge of the Gulf Stream off the southern United States. The distribution of other fishes in the world ocean may be controlled by different variables.

Many nekton seem capable of adjusting easily to rapidly changing physical conditions. Some, however, do not or cannot. For example, occasionally the warm equatorial countercurrent that flows toward the coast of Ecuador penetrates the coastal region as far south as Peru. The sudden influx of warm water (called "El Niño") causes great mortality to marine life and birds that rely upon fish for food.

BENTHOS

Benthic organisms are plants and animals that (1) are *sessile*, or attached to the bottom (sponges, barnacles, corals, oysters, sea anenomes, seaweeds), (2) burrow into the substrate (clams, worms), (3) creep over the sea floor (starfish, crabs, lobsters, snails), and (4) spend much of their lives associated with the bottom, as certain fishes do.

The benthos of nearshore regions is characterized by the profusion and variety of organisms living there. In this environment, wave forces and tidal currents are extremely vigorous; hence, attached forms predominate. The littoral and sublittoral environments have extremely varied physical conditions, bottom types, and habitats, so it is not surprising that there is such a wide variety of littoral and sublittoral benthic organisms. In fact, nearly all major forms of the animal kingdom are represented in these regions.

A combination of several factors causes the abundance of life in the nearshore environment. Strong currents and associated vertical mixing, high nutrient supply from the land, and adequate sunlight reaching the bottom (permitting the growth of plants) are all important. Strong currents are particularly vital for carrying food to sessile organisms that are incapable of moving. Often food is so plentiful in this environment that animal populations are limited only by the scarcity of space or shelter. It is the luxuriant growth of plant material in the nearshore environment that causes such an abundance of animal life.

Like the nekton, littoral and sublittoral benthic organisms have a definite geographical distribution. Distinct fauna can be recognized in arctic, tropical, and antarctic regions. Some organisms, however, do occur in regions transcending the usual geographical boundaries. The geographical distribution of fauna appears to be controlled primarily by water temperature, but other oceanic factors are also important.

Within any beach or nearshore region, plants and animals tend to distribute themselves in zones that parallel the shore. The position along a beach that a given organism prefers is a function of the tides (degree of exposure), substratum, currents, wave forces, temperature, salinity, light penetration, and other physical conditions.

One scheme to describe the characteristic zones on a rocky shore is shown in Fig. 12-6. Each zone is defined by its position relative to the local tidal range and by the presence of certain distinctive organisms. Such classifications have been devised for exposed coasts, protected coasts, and estuaries.

Organisms living at the bottom seaward of the sublittoral zone are termed *deep-sea benthos*. The numbers of deep-sea benthic organisms decrease rapidly with increasing depth. This decrease led early oceanographers to conclude that no life existed below 700 m. This lifeless region

FIGURE 12-6
Zonation of organisms of the rocky shore environment.
(Photograph courtesy of Peter B. Taylor)

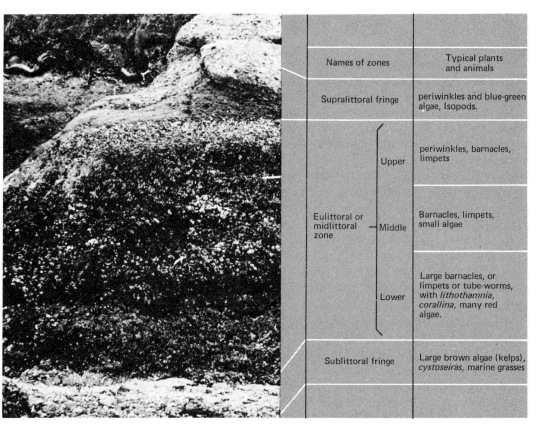

Names of zones		Typical plants and animals
Supralittoral fringe		periwinkles and blue-green algae, Isopods.
Eulittoral or midlittoral zone	Upper	periwinkles, barnacles, limpets
	Middle	Barnacles, limpets, small algae
	Lower	Large barnacles, or limpets or tube-worms, with *lithothamnia*, *corallina*, many red algae.
Sublittoral fringe		Large brown algae (kelps), *cystoseiras*, marine grasses

was called the *azoic zone*. By 1860, however, life was found in the deep sea; since then, thousands of species have been discovered at all depths.

The scarcity of deep-sea benthic organisms is closely related to the decreased availability of food. Abyssal animals are mainly mud-dwellers that scavenge the deep-sea ooze for organic detritus that has settled to the bottom from the surface regions. Consequently, the presence of abyssal fauna is dependent not upon the depth but upon the distance from near-shore areas where organic production is large. The marine sediments farthest from shore (brown clays) contain the least organic detritus, so they support the smallest population in the deep sea.

Generally, deep-sea benthic organisms are widely distributed because of the uniformity of physical and substrate conditions. Nevertheless, there are environmental barriers that limit the migration of organisms—for example, the Wyville Thompson Ridge that separates the North Atlantic and the Norwegian seas. A temperature difference of only 4 degrees (C) across the ridge inhibits migration to the extent that only 11 per cent of the fauna are common to both sides.

READING LIST

BARRINGTON, E.J.W., *Invertebrate Structure and Function*. Boston: Houghton Mifflin Company, 1967. 549p.

MACGINITIE, G.E., AND N. MACGINITIE, *Natural History of Marine Animals*. New York: McGraw-Hill Book Company, 1949. 473p.

MEGLITSCH, P.A., *Invertebrate Zoology*. London: Oxford University Press, 1967. 961p.

RICKETTS, E.F., J. CALVIN, AND J.W. HEDGPETH, *Between Pacific Tides* (4th ed.) Stanford, Calif.: Stanford University Press, 1968. 614p.

SOUTHWARD, A.J., *Life on the Seashore*. Cambridge: Harvard University Press, 1965. 153p.

YONGE, C.M., *The Seashore*. London: Collins, 1949. 311p.

ONE
TWO
THREE
FOUR
FIVE
SIX
SEVEN
EIGHT
NINE
TEN
ELEVEN
TWELVE
THIRTEEN
FOURTEEN
FIFTEEN

marine ecology

Marine ecology is the study of interrelationships among the organisms of the sea and the relation of these marine organisms to their environment. This definition is so broad that much of marine biology must be considered a part of marine *ecology*.

In this chapter, both aspects of marine ecology are presented. First, we discuss how marine organisms are adapted to environmental factors in the ocean. Secondly, we describe how the feeding habits of marine organisms show their interrelationships.

13.1 ADAPTATIONS OF PLANT AND ANIMAL GROUPS TO THE MARINE ENVIRONMENT

Life has been found virtually everywhere in the sea. The myriad of species of marine organisms has survived, because each species has become adapted to its environment. These adaptations involve all aspects of an organism's life, including its structure, reproductive processes, physiology, feeding habits, and movement. Each organism possesses many complex and subtle features that represent its evolutionary adaptations. A thor-

226

ough discussion of so vast a subject is beyond the scope of this text, so we describe here only the broad classes of organisms in the sea. Examples are given of obvious adaptations that contribute to the success of the organisms in each group.

PHYTOPLANKTON

Marine plants, like all plants, require light. They flourish, therefore, only if they remain suspended in the photic zone sufficiently long enough to grow and reproduce. The need to remain continually in the surface water of the ocean has led to several adaptations. Marine plants have little or no ability to swim. Their protoplasm is usually denser than seawater (1.02 to 1.06 g per cu cm), so they have developed ways to retard their sinking rate. These adaptations greatly influence the size, structure, and density of plants in the ocean today.

SIZE. The rate of sinking of an object is determined by its weight and its surface area. An object with a large surface area will settle through water more slowly than an object with the same weight but with a smaller surface area. The surface area relates to frictional retardation, and the volume is proportional to the weight of the object. An object with a large ratio between surface area and volume will sink more slowly than one equally dense but with a smaller ratio. The magnitude of the ratio is determined by size of an object; that is, the larger the object, the smaller the ratio, as shown by the following equation:

$$\text{ratio} = \frac{\text{area}}{\text{volume}} = \frac{4\pi r^2}{(4\pi r^3/3)} = \frac{3}{r} \qquad [13\text{-}1]$$

Two equally dense spheres with radii of 1 mm and of 0.01 mm have ratios of 3 and 300, respectively. Thus, other factors being equal, the smaller the sphere is, the slower the sinking rate.

The ratio between surface area and volume has one other important aspect. Phytoplankton obtain nutrition by passing nutrients through their surfaces. Thus, it is advantageous to have a large surface area relative to the quantity of protoplasm to be nourished (i.e., a large surface area-volume ratio).

STRUCTURE. Some of the phytoplankton have shapes and appendages that cause them to sink in nonstraight paths (Plate II). Disc-shaped diatoms, for example, sink in a zig-zag path; needle-shaped forms follow a wide circular path. In both cases, the organisms sink less rapidly than the theoretical sinking rate calculated on the basis of the mass of the organism.

Appendages or spines retard sinking by increasing the surface area of the organism. The spines on some dinoflagellates are asymmetrical in order to keep the organism oriented in such a way that its flagella can move it toward the surface.

The relative length of appendages can be related to water temperature. In warmer waters, where viscosity is low, appendages and spines are longer and more elaborate than in colder waters, which have higher viscosity. This adaptation takes advantage of the retarding effect of viscosity upon the sinking rate of an object.

DENSITY. Phytoplankton adaptations to the marine environment include the constitution of the cell. Many species contain oil sacs that decrease their density. Accumulations of oil are also useful as food reservoirs while an organism awaits more favorable environmental conditions. Seasonal differences in density are observed in some species; that is, by accumulating oil, summer forms become less dense than their winter counterparts. This adjustment in density accompanies changes in water viscosity in such a way that the sinking rate does not increase.

ZOOPLANKTON

Many zooplankton species depend upon marine plants for food. Because they must remain close to the phytoplankton, many adaptations similar to those of the phytoplankton are observed. Zooplankton density is minimized by maintaining a high water content and by having oil sacs in their body and eggs. The surface area-weight ratio is increased by small size and various appendages. The copepods, for example, possess bristles, plumes, spines, and antennae, all of which increase the body surface area and swimming ability (Fig. 13-1).

MOBILITY. The zooplankton employ various means of locomotion. Copepods and euphausids swim by using muscular action. Ctenophores and marine worms have ciliary motion. Jellyfish move by jet action. Organisms can float by means of a buoyant froth, as does the barnacle, or by gas-filled floats, like the jellyfish *Physalia*. The jellyfish *Velella* is nicknamed "by-the-wind-sailor" because of its characteristic sail (Fig. 13-2).

DAILY MIGRATION. An interesting environmental adaptation of pelagic organisms, especially the zooplankton, is the occurrence of daily vertical migrations. Investigations have shown that many of the pelagic zooplankton make daily journeys between the surface and deeper water. Although these organisms are classed as plankton, they are not completely passive, and they continually change their position in the water column. Small plankton crustaceans can swim at rates of 16 m per hr; larger species are able to swim 95 m per hr.

The diagram in Fig. 13-3 traces the vertical distributions of the adult copepod *Calanus* throughout a 24-hour period. During the brightest part of the day, the vertical pattern of organisms is approximately diamond-shaped; the greatest numbers occur at a depth of 15 m. The organisms

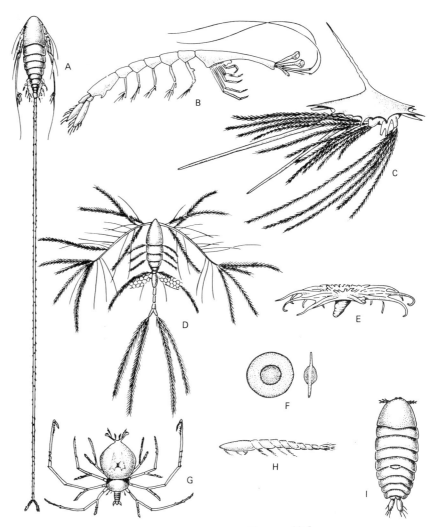

FIGURE 13-1

Some floating adaptations of plankton animals: (a) copepod (*Aegisthus*);
(b) decapod (*Lucifer*); (c) barnacle nauplius; (d) copepod (*Oithona*);
(e) holothurian (*Pelagiothuria*); (f) pelagic egg of copepod (*Tortanus*);
(g) phyllosoma larva of lobster; (h, i) copepod (*Sapphirina*),
side and dorsal views. (After Sverdrup, Johnson, & Fleming; 1942;
by permission of Prentice-Hall, Inc.)

begin migrating upward in late afternoon and become grouped at the surface. They are presumably feeding. Downward migration begins after
midnight and is completed at noon, thus ending the daily cycle. The
entire migration may extend over several hundred meters of depth. However, the vertical range shown in this figure is probably a more characteristic value.

Observations reveal that the greatest number of individuals are
found at a level where light and other factors are optimum. It is not
known whether the light stimulates other responses (hunger or preference

FIGURE 13-2

The jellyfish *Vellela sp.* floats at the sea surface and has a sail to
provide a certain degree of mobility. Photograph approximately
natural size. (Photograph courtesy of Dora P. Henry).

for a particular temperature). However, light appears to be the most
important motivation; the depth of migration in winter and in summer
differs as light penetration differs. In addition, the portion of the population found at a single level may consist of individuals of uniform age or
sex, thus suggesting that there is a different migration pattern for each
species, sex, and age. The net result is a more or less continuous, but
uneven, pattern of vertical migration for the total population.

A seasonal migration pattern is also observed in certain species of

FIGURE 13-3

Vertical distribution of the adult female copepod, *Calanus finmarchicus*
during a 24-hour interval. (After Nicholls; on the Biology of Calanus;
which appeared in the Journal of the Marine Biological Association,
1933, by permission of the Cambridge University Press.)

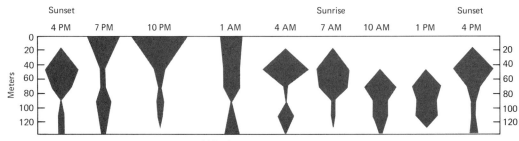

width of pattern represents relative
abundance at each depth

the same copepod *Calanus*. At the onset of winter, they take on food reserves and sink to depths of 200 to 300 m to spend the winter months. An upward migration begins about February, and the organisms arrive in surface waters during the spring phytoplankton bloom. This cycle appears to be an adaptation to the scarcity of food (phytoplankton) at the ocean surface during winter.

BENTHOS

LITTORAL AND INNER SUBLITTORAL ORGANISMS. To succeed in this environment, an organism must be adapted for life under the most variable marine conditions. Temperature and salinity fluctuations of significant proportions occur on a seasonal basis. In shallow regions, such fluctuations also occur daily. The substratum may change from season to season. In addition, current and wave forces are extremely strong. Here, the problems are not food but the competition for shelter and space and the struggle against desiccation in the intertidal zone.

A prime requisite for organisms in this habitat is the physiological ability to withstand significant temperature and salinity variations. Some neritic (shallow-water) phytoplankton species form resting spores or dormant stages. This adaptation allows them to avoid adjusting to the rigors of the sublittoral environment. Other organisms have developed tolerances, becoming *eurythermal* and *euryhaline*; that is, they can withstand large temperature and salinity fluctuations.

Modifications in structure and mode of living are particularly noticeable in the organisms of the nearshore benthos. Sessile, or attached, forms predominate; they are flattened, massive, and streamlined to withstand strong currents and wave forces. These organisms must feed upon food suspended in the water. In the nearshore environment, therefore, filter feeding (straining organic particles from the water) is a highly successful adaptation, because strong currents assure a generous supply of suspended particulate matter and plankton.

The nearshore mobile forms are also sturdy and compact. Some organisms, such as sea urchins and pelecypods, can burrow into rocks; others, especially certain fish and mollusks, rely upon sucking discs to attain a degree of attachment. Barnacles attach themselves to surfaces by cementation. Mussels attach themselves with byssus threads that they secrete for this purpose. By using mucus as adhesive, certain worms cement sand grains into tubular shelters attached to rocks.

Feeding habits of these mobile organisms vary however. Many, like pelecypods, are filter feeders. Others, such as some worms, sea urchins, pelecypods, and crustaceans, are deposit feeders that scavenge organic debris from rocks, sand, or mud. Predation occurs among the shore fish, crabs, birds, and starfish.

In the intertidal region, desiccation is a major problem, but there are a variety of ways for organisms to survive. Some bury themselves in

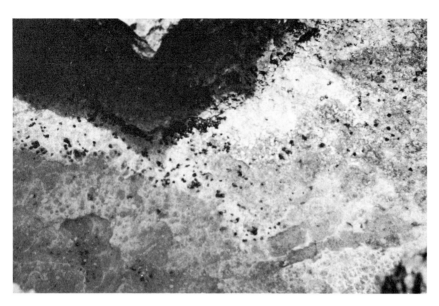

Figure 13-4
When exposed at low tide the small *Littorinid* snails tend to
cluster to retard desiccation. Those snails still underwater are dispersed
while those exposed are clustered.

the substrate; others live in tide pools that retain some water at low tide.
Many forms (i.e., barnacles, snails, small mussels) protect themselves from
exposure by retracting into their shells. Littorinid snails form clusters to
retain moisture (Fig. 13-4). Other organisms actually do dry out somewhat but replace this loss of water by absorbing more water during the
next high tide. Sea urchins and sea anemones cover themselves with shell
fragments while exposed (Plate IV C), although some sea anemones produce a mucous secretion that inhibits desiccation.

The brown algae that inhabit the inner sublittoral are especially
well adapted to the rigors of the nearshore environment. To keep from
being washed ashore by large waves, these plants attach themselves by a
holdfast to rocks on the bottom. Large gas bladders float the fronds to the
surface region where incident radiation is abundant (Fig. 11-2). These
plants may grow to 80 m in length and produce thick kelp forests seaward
of the surf zone in many coastal areas.

DEEP-SEA ORGANISMS. There is extreme uniformity in the environment of the deep sea, so organisms have not adapted (or have lost the
ability) to tolerate wide differences in temperature and salinity. Consequently, animal and plant life in this region is *stenothermal* and *stenohaline*. An example of this lack of tolerance is the faunal assemblages on
either side of the Wyville Thompson Ridge in the North Atlantic (noted
in Chapter 12).

232

Organisms of the deep sea are not streamlined, and their skeletons are relatively frail. These characteristics reflect the lack of strong currents and wave forces in the environment. The absence of strong water motions and the low concentrations of suspended matter explain why deep-sea organisms are deposit feeders that must ingest large quantities of deep-sea sediment and digest whatever organic matter is available.

NEKTON

FISHES. Most fishes are predators that feed on highly mobile organisms. The most obvious adaptations of fishes are those that provide advantages in movement. They have streamlined shapes, and some forms are covered with a slimy substance to decrease frictional resistance. Their muscles, nervous system, and sense of vision are extremely well developed.

FIGURE 13-5

Deep-sea fishes: (a) *Macropharynx longicaudatus* (after Brauer), length 15.1 cm, from 3500 m depth; (b) *Bathypterois longicauda* (after Günther), length 7.6 cm, from 550 m depth; (c) *Gigantactis macronema* (after Regan), length 13.3 cm, from 2500 m depth; (d) *Linophryne macrodon* (after Regan), length 5.3 cm, from 1500 m depth; (e) *Malacostus indicus* (after Brauer), length 8 cm. from 914–2500 m depth. (After Sverdrup Johnson, & Fleming, 1942, by permission of Prentice-Hall, Inc.)

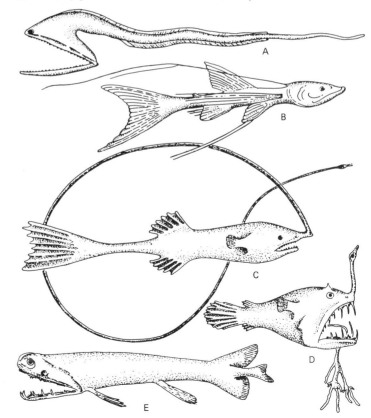

These adaptations permit fishes to expend minimal energy while hunting their prey. This ability is important, because the energy intake from food must exceed the energy used in obtaining food or the organism dies.

For protection from predators, fishes are camouflaged. They are generally blue or gray on top and light-colored underneath so that they blend with their backgrounds.

WHALES. Considered to be the most specialized of all mammals, whales exhibit some interesting adaptations. First, their large size and weight (100,000 kg) assures that they have few enemies. Secondly, several species of whales have developed a filtering mechanism made of baleen in their upper jaw and feed on planktonic animals (Plate IX E). This adaptation is advantageous, because it does not require the expenditure of energy needed to chase fast-swimming fish. Third, whales have apparently descended from terrestrial mammals; but in the evolutionary transition, the hind legs were lost and the forelegs were modified into fins for steering during swimming. Finally, a thick layer of blubber insulates these warm-blooded animals from cold waters in the polar seas and at depth elsewhere.

Whales must breathe air at the sea surface; but they can dive to depths exceeding 1000 m and can remain submerged for as long as an hour. This feat is made possible by the whale's ability to discontinue peripheral circulation during dives so that oxygenated blood is used only to maintain the vital organs.

The growth rate of young whales is unique in the animal kingdom. A newborn calf can be 7 m long and weigh 2000 kg. Some species mature in two years. By then, the calf has increased its weight over 30,000 kg per year.

BATHYPELAGIC ORGANISMS. The organisms living at depths greater than 1000 m include crustaceans, chaetognaths, and fishes. Generally, the fishes are small (10 to 20 cm) and are colored black, black-violet, or brownish. Life must be carried on in darkness, so these organisms possess many adaptations to assist them in food-gathering, reproduction, and protection.

Food-procuring devices are the most evident. Some forms have immense mouths in relation to their size, possess formidable teeth, and have elastic stomachs and abdominal walls. They can thus swallow organisms three times their size (Fig. 13-5) and enjoy a wide range in diet. In another case of adaptation, the angler fish carries its own luminous lure, which extends four times its body length, presumably for food gathering. Some lures are equipped with spines to decrease the chance of being eaten by other predators. Male angler fish that live in deep water (2000 m) are dwarfed and are permanently attached to the females as parasites (Fig. 12-4). This bizarre adaptation assures the fertilization of eggs in an environment where the population is highly dispersed, and the search for a mate might require the expenditure of too much energy.

The eyes of bathypelagic organisms show several kinds of adaptation. Some near-bottom forms are blind or have degenerate eyes, but others have well-developed eyes. Some forms have enlarged eyes to receive the faint rays of light that reach mid-depths. For the same reason, other fishes show an increase in visual rods in the retina and even some degree of telescopic binocular vision.

Many forms of the bathypelagic fish are *bioluminescent* and have their sides lined with *photophores*, which is a light-producing mechanism. Although this mechanism is found in organisms at all depths, its advantages and uses are not definitely known. It might serve as a lure for prey, a warning for predators, a signal for the opposite sex at spawning time, or a signal for establishing territoriality among members of the same sex.

13.2   THE MARINE FOOD WEB

The equilibrium in food energy in the ocean is visualized as a *food web*, which is a complex network of paths of food energy transfer. Each path is termed a *food chain*. Every major transferal of food energy defines a *trophic level* in the food chain. In the first level of the food web, plants manufacture organic matter. This level is called *primary production*, because solar energy and inorganic nutrients are combined into the organic matter that forms the ultimate source of food for all organisms in the sea.

THE FIRST TROPHIC LEVEL

Plants complete the first step in the food web by converting inorganic nutrients into organic matter. Plants contain chlorophyll that allows them to use radiant energy from sunlight to combine carbon dioxide and water to produce carbohydrate (CH_2O) and oxygen. This process is called *photosynthesis* and is stated in Eq. [13-2]. The reaction proceeds in the direction of the arrow.

$$CO_2 + H_2O \xrightarrow{\text{solar energy}} CH_2O + O_2 \qquad [13\text{-}2]$$

In this process, radiant energy from the sun is stored when low-energy inorganic compounds (CO_2 and H_2O) are converted into high-energy organic compounds (carbohydrates). Only plants can accomplish this primary production of organic matter.

All organisms obtain the energy to carry on life processes by oxidizing organic matter, a process called *respiration*. Respiration is the reverse of photosynthesis, and it proceeds according to the equation:

$$CH_2O + O_2 \longrightarrow CO_2 + H_2O + \text{energy as heat} \qquad [13\text{-}3]$$

Carbohydrate is oxidized to CO_2 and H_2O with a release of thermal energy equivalent to the amount required to produce the sugar. When 180 g of glucose are oxidized, 674 kcal of heat energy are released. Organisms always respire, so growth can occur only if the net photosynthetic energy produced during sunlight hours exceeds the energy lost by respiration during both night and day.

Once organic molecules are produced by photosynthesis, they can be used in several ways. Carbohydrate not consumed in respiration can be (1) converted into other forms of carbohydrate, such as glucose or cellulose for cell walls or starch for reserve food; (2) converted to lipids or fatty substances; or (3) combined with inorganic nitrogen and phosphorus compounds to form protein. These organic compounds are used by the cell to form the live substance, *protoplasm*. The entire process is diagrammed in Fig. 13-6.

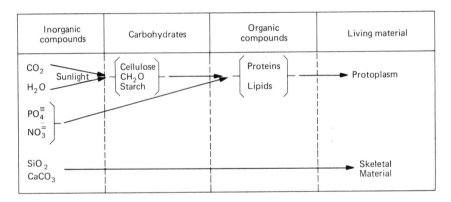

FIGURE 13-6
Various steps in the
production of living material.

The chemical analysis of dried phytoplankton indicates that it has the following components: carbon, 45 to 55 per cent by weight; nitrogen, 4.5 to 9 per cent; phosphorus, 0.6 to 3.3 per cent; varying amounts of silicon (up to 25 per cent) or calcium carbonate (up to 25 per cent), depending on the organism; and various minor elements (potassium, sodium, magnesium, calcium, and sulfur). Although these percentages vary considerably, according to the organism and the state of nutrition, they do indicate the importance of the nutrient elements to the basic productivity in the sea.

In conclusion, three important deductions can be made:

1. Plants must remain in the sunlit regions of the ocean waters (i.e., the surface region) a significant portion of the time if they are to flourish.
2. During the growing season, a decrease in the nutrient concentration in stratified surface water should accompany an increase in plant growth.
3. The oxygen concentration in the surface water should increase with increased productivity if other factors remain approximately constant.

DECOMPOSITION OF ORGANIC MATTER

The decomposition of waste and dead organic matter is a necessary process in the sea. It is carried out by bacteria, which convert organic compounds formed at all trophic levels into inorganic nutrients that replenish those consumed by plants at the first trophic level.

Bacteria are microscopic, one-celled organisms that are found throughout the ocean. They are usually attached to the surface of organic debris or to other organisms, or they may live on the sea floor. Freely suspended bacteria can be found in seawater; however, their concentration is small (on the order of 10 per cu cm).

In the presence of dissolved oxygen, bacterial remineralization proceeds as described by Eqs. [13-4] and [13-5].

$$CH_2O + O_2 \longrightarrow CO_2 + H_2O \qquad [13\text{-}4]$$

$$NH_3 + 2O_2 \longrightarrow H^+ + NO_3^- + H_2O \qquad [13\text{-}5]$$

Equation [13-4] represents the breakdown of carbohydrate by aerobic bacterial metabolism. Ammonia, NH_3, is formed by the bacterial breakdown of organic nitrogen compounds; it is also present in the excretions of zooplankton. Equation [13-5] describes the oxidation of ammonia to an inorganic nitrate nutrient. Notice that remineralization of organic matter consumes oxygen.

Remineralization completes the basic food cycle, which is shown in Fig. 13-7. In the photic zone, plants convert nutrients to organic matter. This material either sinks or is eaten by animals and carried below the photic zone. In the aphotic zone, bacteria eventually convert it back to nutrients.   Nutrients are returned to the photic zone by the physical processes of diffusion and turbulent mixing.   Where this process is intensified by upwelling, as along the west coast of the United States and the coast of Peru, biological production reaches some of the highest levels in the world ocean.

In basins where vertical circulation is poor, the decomposition of organic debris slowly removes oxygen from the bottom water until the processes described by Eqs. [13-4] and [13-5] can no longer occur and *anaerobic* decomposition prevails.   This kind of decomposition involves the processes of *denitrification* and *sulfate reduction* (see Chapter 4).   Equations for denitrification are:

$$5CH_2O + 4H^+ + 4NO_3^- \longrightarrow 2N_2 + 5CO_2 + 7H_2O \qquad [13\text{-}6]$$

$$5NH_3 + 3H^+ + 3NO_3^- \longrightarrow 4N_2 + 9H_2O \qquad [13\text{-}7]$$

and for sulfate reduction:

$$2CH_2O + 2H^+ + SO_4^= \longrightarrow H_2S + 2CO_2 + 2H_2O \qquad [13\text{-}8]$$

$$NH_3 + 2H^+ + SO_4^= \longrightarrow H^+ + NO_3^- + H_2S + H_2O \qquad [13\text{-}9]$$

Note that the result of these reactions are the inorganic compounds, CO_2, H_2O, H_2S, and free nitrogen (N_2).   Hydrogen sulfide (H_2S), produced by

FIGURE 13-7
The first cycle in the food web.

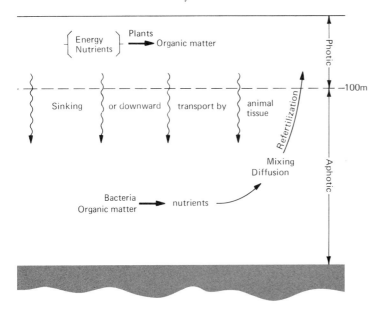

TABLE 13-1

VARIETY OF LENGTH OF FOOD CHAINS

Plants→copepods→juvenile herring→chaetognath
→adult herring→salmon→bacteria
Plants→zooplankton→whale→bacteria
Plants→zooplankton→bacteria

sulfate reduction, is toxic to most organisms. It accounts for the bad odor and lack of life in the deep portions of such stagnant basins as the Black Sea and many fjords.

The remineralization reactions complete the cycle of the generation and consumption of inorganic nutrients in the sea. Here we have considered only the first trophic level. To appreciate the cycle, we must see how other trophic levels are involved.

HIGHER LEVELS IN THE FOOD WEB

Animals cannot appreciably synthesize organic matter from inorganic substances. They must, therefore, rely on plants and other animals for their food. The primary grazers of marine plants are the zooplankton. The copepods are the largest single group of grazers. Their abundance and distribution are determined by the abundance and distribution of phytoplankton. The food relationships of the other marine organisms are much more complicated. The relationships involving the herring, for example, are illustrated in Fig. 13-8. This diagram shows how animals from several trophic levels graze on phytoplankton and that an animal can occupy several trophic levels during its development. The figure also shows that food chains are of various lengths. Some are short and involve only one or two transfers of food from phytoplankton to bacteria; others, like those containing the herring, involve several transfers. Some examples are given in Table 13-1.

Figure 13-7 also shows how each individual in a high trophic level (e.g., a herring) must rely on several individuals from a lower level for food. Usually, an organism uses about 10 per cent of the protoplasm that it consumes for the production of tissue; the rest it either uses for maintenance or excretes. As a result, there is a 90 per cent loss of food energy at each trophic level. An organism such as the herring is dependent upon a huge number of phytoplankton for its growth. If this food chain is four levels long, a metric ton of herring would consume 1,000 metric tons of phytoplankton, as illustrated by the following diagram:

1,000,000 kg ⟶ 100,000 kg ⟶ 10,000 kg ⟶ 1,000 kg (1 metric ton)
phytoplankton   zooplankton   crustaceans   herring

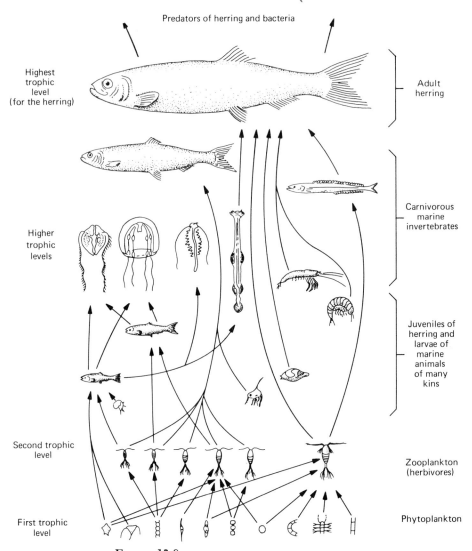

Predators of herring and bacteria

Highest
trophic
level
(for the herring)

Adult
herring

Higher
trophic
levels

Carnivorous
marine
invertebrates

Juveniles of
herring and
larvae of
marine
animals
of many
kins

Second trophic
level

Zooplankton
(herbivores)

First trophic
level

Phytoplankton

FIGURE 13-8

Food web of the herring at various periods of life.
(After Hedgpeth, J. W., *Concepts of Marine Ecology;* Geological Society of America Memoir 67,
1957.)

The primary production of phytoplankton in the world ocean is estimated to be 135×10^{12} kg per year. This figure is close to the annual production of plants on land. The annual harvest of fish is 50×10^9 kg, although the potential yield could reach 180×10^9 kg, which is equivalent to a three-step food chain.

Only a small portion of the net production of the sea—that is, the fish—is readily available to man. Fishes, however, are in the higher trophic levels. Compared to the plankton at the lower levels, fishes have small

240

populations and slow reproductive rates. They are captured for food by man because they yield the most food for the least expenditure of energy. An improved efficiency in obtaining food from the sea is possible if fisheries with a shorter food chain are developed. For every trophic level eliminated, 7 to 10 times more protein is available from the plant material formed at the first trophic level. Thus, it is much more efficient to develop and maintain the sardine, anchovy, and mackeral fisheries than to rely on tuna or cod to feed the world. The Peru anchovy fish industry, the largest catch of an individual species in the world, is a good example of a short food chain with high productivity. It would be still better to use zooplankton for our basic source of protein, but they are so difficult to catch that the effort would be out of proportion with the return.

13.3 THE ANNUAL CYCLE OF PHYTOPLANKTON

At the first two trophic levels, there is a tendency for supply and demand in the food chain to reach a balance that maintains fairly constant amounts of phytoplankton and zooplankton at any part of the ocean. The balance is easily upset, however, because of changes caused by other environmental factors. The most significant changes occur seasonally, resulting in cyclic abundance of phytoplankton. The environmental factors causing this annual cycle are: (1) the amount of incident sunlight, (2) the concentration and supply of nutrients in the surface water, (3) the thickness of the layer of surface water mixed by the wind, and (4) the abundance of zooplankton that graze upon phytoplankton.

GENERAL ENVIRONMENTAL FACTORS

SUNLIGHT. The amount of solar radiation striking the sea surface depends upon the season and upon geographic latitude. The extent of seasonal variations in irradiation also depends upon the latitude. In the ocean at high latitude, seasonal variation in sunlight is extreme; and so phytoplankton are able to grow for only a short period. In the tropics, there is always ample sunlight for phytoplankton growth because seasonal changes in solar radiation are minimal.

NUTRIENTS. These substances are consumed from seawater in the photic zone as phytoplankton growth proceeds. If nutrients are not replenished from below the pycnocline, where most of the bacterial remineralization takes place, the nutrients become depleted, and phytoplankton growth is limited. Replenishment occurs by molecular diffusion from below and, most important, by turbulent mixing.

MIXED LAYER. During storms, wind-generated waves produce turbulence in surface water. The thickness of the layer of turbulent mixing depends upon the frequency and intensity of storms and the degree of stability of the water column. Both storms and differences in density stratification occur seasonally, so the thickness of the mixed layer also varies seasonally. Phytoplankton are kept in suspension by the turbulence in the mixed layer. If the turbulence is so intense that mixing extends below the photic zone, phytoplankton will be swept down to a region of insufficient sunlight for a period long enough to kill them. On the other hand, such intense mixing will bring remineralized nutrients from below into the photic zone.

GRAZING BY ZOOPLANKTON. A seasonal increase in the abundance of phytoplankton increases the food supply for grazers, in turn, leading to seasonal increases in the zooplankton population. As grazing increases, the population of phytoplankton is reduced and less food is available to the zooplankton.

The seasonal variations of the environment produce a corresponding imbalance between phytoplankton and grazing zooplankton. The shifts in the equilibrium of the plankton community are represented by the yearly cycle of diatoms shown in Fig. 13-9. However, the pattern of phytoplankton growth is not uniform over the entire ocean because the environmental factors and the way they are disrupted vary with location.

ENVIRONMENTAL FACTORS IN THE OPEN OCEAN (OCEANIC ENVIRONMENT)

TROPICAL REGIONS. Low-latitude portions of the world ocean (between 30 degrees N and 30 degrees S latitude) are characterized by high-incident solar radiation and a strong, permanent pycnocline. There is adequate sunlight throughout the year; but strong vertical mixing does not occur, and so the mixed layer cannot be refertilized. As a result, plant production in this region remains at a low level throughout the year.

The only variation in this pattern is at the equatorial divergence (see Chapter 7) where nutrient-rich subsurface water wells up into the sunlight (Fig. 7-12). This water supports 10 times the plant production of the adjacent tropical water.

TEMPERATE REGIONS. Between 30 degrees and 55 degrees latitude, the productivity may vary as shown in Fig. 13-9. The phytoplankton population is smallest in the winter, because the light is insufficient and the mixed layer is deep. In high latitudes, the seasonal pycnocline is destroyed as temperatures decrease and winter storms drive vertical circulation deep into the water column. The surface water becomes fertilized by mixing with the nutrient-rich water below.

In the spring, several important changes occur. Illumination increases, winter storms subside, and the weather becomes warmer. With

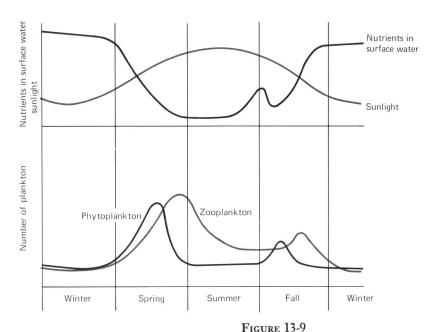

FIGURE 13-9
A schematic diagram of yearly phytoplankton and zooplankton
production occurring in temperate latitudes. Yearly changes
in nutrient concentration and sunlight are included also.

the advent of a stratified water column, more sunlight, and more abundant
nutrients, phytoplankton proliferate in a spring bloom.

During the summer season, the marine plant population decreases
again. Nutrient-depleted surface water cannot be refertilized, because the
highly stratified water cannot be mixed. Grazing reduces the population.
Even though sunlight is at a maximum, the population cannot regenerate
itself because of the deficiency of nutrients.

In the autumn, stability decreases as the surface water cools. If an
isolated autumn storm stirs the surface but does not completely destroy the
seasonal pycnocline, the surface water can be refertilized. Daily illumination, although decreasing, is still sufficient to promote an autumn bloom.
If autumn storms continue and are strong, mixing extends below the
photic zone and the bloom is deterred. The occurrence and magnitude of
autumn blooms vary from year to year, depending on the weather.

POLAR REGIONS. The growth pattern at high latitudes is similar to
that in temperate regions but more variable. The growing season is
shorter, and storms are more frequent and more intense. A spring bloom
occurs after the ice pack melts. Often there is no fall bloom because of
adverse climatic conditions.

Under the permanent ice pack, there is but one bloom per year.
During July and August, the snow cover disappears, and sufficient sunlight
penetrates the ice to allow plant growth.

ENVIRONMENTAL FACTORS ON THE CONTINENTAL MARGINS (NERITIC ENVIRONMENT)

The annual pattern of productivity on the continental shelf is unique. Nearly all of the water column is penetrated by sunlight, and nutrients remineralized by bacteria on the seabed are accessible when mixing occurs. As a result, nutrient depletion is not nearly as critical as in the open ocean, and productivity is considerably higher.

In tropical regions, productivity is high throughout the year. In temperate regions, the spring bloom is produced by increased sunlight and accumulation of winter nutrients. A significant summer bloom occurs when bacterial regeneration increases. In addition, coastal upwelling, caused by favorable winds (Chapter 7), produces extremely large productivity peaks. The daily productivity in some coastal upwelling areas is 100 times that in the tropical regions of the open ocean.

13.4 SOME THEORETICAL ASPECTS OF FOOD-CHAIN RELATIONSHIPS

Biological oceanographers are interested in knowing precisely how the variables associated with biological productivity in the sea interact.

FIGURE 13-10

Comparison of observed seasonal cycles of phytoplankton (solid lines) with theoretical cycles (shaded lines). (Courtesy of the Journal of Marine Research).

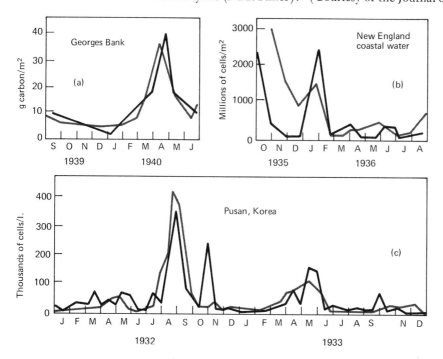

If we relate organisms and environmental factors on a theoretical framework, we may be able to predict the phytoplankton (or zooplankton) population at any given time. Several extremely simplified theoretical models have been proposed; even the simpler models have depicted the changes observed in phytoplankton population during the seasons.

One example of a simple model is given in Eq. [13-10]:

$$\text{Rate of change of population size} = (P_h - R - G) \text{ population size} \qquad [13\text{-}10]$$

This equation states that the rate of change in the total phytoplankton population at any instant is proportional to the size of the phytoplankton population at that instant. The factor of proportionality is the coefficient of photosynthesis (P_h) minus a coefficient of respiration (R) minus a coefficient of grazing (G). $P_h - R$ is a measure of the net biological growth of the population and is related to the incident of radiation, transparency of the water, nutrient supply, depth of the mixed layer, and the water temperature. The grazing coefficient (G) is related to the herbivore population.

Phytoplankton populations monitored on a seasonal basis in several areas have been compared with the theoretical predictions, and the results are surprisingly good (Fig. 13-10). Indeed, with successful theoretical models, we can extend our understanding of ecological relations beyond the facts obtained by observation alone. Models of a more complex nature are being investigated by oceanographers, and they represent an important part of the study of marine ecology.

READING LIST

DOWDESWELL, W.H., *Animal Ecology* 2nd Edition. New York: Harper, 1961. 209p.

HEDGPETH, J.W., AND H.S. LADD, *Treatise on Marine Ecology and Paleoecology*, Geological Society of America Memoir 67. New York, 1957. 2 vols.

JORGENSEN, C.B., *Biology of Suspension Feeding*. New York: Pergamon Press, 1966. 357p.

MOORE, H.B., *Marine Ecology*. New York: John Wiley & Sons, Inc., 1958. 493p.

WILSON, D.P., *Life of the Shore and Shallow Sea*. London: Ivor Nicholson & Watson, Ltd., 1935. 150p.

ONE
TWO
THREE
FOUR
FIVE
SIX
SEVEN
EIGHT
NINE
TEN
ELEVEN
TWELVE
THIRTEEN
FOURTEEN
FIFTEEN

marine sediments

Layers of sediments at the bottom of the sea reflect physical, chemical, geological, and biological differences in the world ocean. Geographic factors are also evident in the diversity of sediments that exists on the bottom, but this remarkable diversity usually goes unnoticed, because it is hidden beneath the water. There are other evidences of diversity however. Everyone knows that beaches are rocky or sandy and that tidal flats are usually a sticky mud. A careful observer may also notice that the sand on the wet portion of some broad beaches is quite firm and can easily support an automobile (Fig. 14-1). Under the water, however, especially in the surf, the same sand is quite loose and yields under the weight of a person. Children are quick to notice this fact and make a sport of twisting their feet to bury them in the sand.

In addition, the coarseness and color of sand are frequently different on adjacent beaches in the same area. There are, of course, differences in sands on the various coasts of the world ocean. Variations in the grain size and color of the sediments also occur offshore. The sediments in the ocean depths are clay-like and very sticky at the sea floor itself but somewhat firmer below.

The most obvious feature of marine sediments is their bulk prop-

246

FIGURE 14-1
Photo of cars on the beach at Daytona, Florida. (From Daytona Chamber of Commerce).

erties. Beside color and texture, the degree of *compaction* is significant. Sands are either loose-packed or close-packed, depending upon whether they are stirred by turbulence in the surf or moved to and fro by waves washing on the beach. Clay-like sediments become more compact as they are buried and as *interstitial* water is expelled by the weight of overlying particles. If we study only the bulk properties of marine sediments, however, we usually do not find an explanation for the diversity of marine sediments. Consequently, it is necessary to analyze (i.e., separate into component parts) the sediments of the ocean bottom. Once this procedure is accomplished, each component can be examined to discover what a marine sediment is, why it appears as it does, and why it occurs where it is found.

247

14.1 THE COMPONENTS OF MARINE SEDIMENTS AND THEIR DISTRIBUTION

Examination of the sediments on the ocean bottom reveals that they are composed of a variety of solid materials that can be grouped into five categories on the basis of their sources. These categories, or components, are: *terrigenous, biological, extraterrestrial (cosmic), submarine volcanic,* and *chemical.* The number and kinds of components represented in a particular sediment depend upon the location of that sediment in the world ocean, so they are indicative of the mode of origin of the sediment. Terrigenous and biological components are by far the most abundant in marine sediment. However, the other sediment-producing components are important locally.

TERRIGENOUS COMPONENTS

Terrigenous components are produced by the weathering and erosion of rocks.* These rock remnants are transported to the ocean by rivers, wind, and ice. Most of this material is ultimately derived from continental masses by the processes of chemical and physical weathering. During weathering, each of the rock-forming minerals behaves somewhat differently. Calcite dissolves almost totally. Ferromagnesian minerals decompose and yield quartz, clays, and dissolved materials. Mica and feldspar behave similarly, except that mica reacts a bit slower and feldspar slower still. Quartz is unaffected except for slight solution. Most accessory minerals are practically unaffected. When weathering has been virtually completed, the rock is reduced to (1) quartz particles in sizes from 1 mm to 0.05 mm in diameter, (2) *clay* particles (aluminosilicates) that are as small as 0.002 m in diameter, (3) accessory minerals in sizes similar to quartz, and (4) aqueous solution.

The solid products of weathering are smashed and ground into finer sizes as they are transported by rivers. In addition, winds may blow the finer materials directly into the ocean. At high latitudes, soil and rock fragments may freeze in the glaciers, which then move this material to the ocean. The combined destruction and removal of the original rock is termed *erosion.* In some cases, however, there may be no weathering and erosion at all. For example, when volcanoes erupt on land, they discharge rock, dust, and ash directly into a wind that carries it to the ocean.

The products of all these processes form the terrestrial components of marine sediments. They make their way to the ocean bottom either as a shower of discrete particles, or transported by waves and currents, or as part of a *turbidity flow,* which is a submarine "river" of sedimentary materials flowing from the site of a submarine slide or originating in some other manner (for example, from flooded rivers).

---

* Definitions and descriptions of important rocks and minerals are given in the Appendix.

TABLE 14-1

SIZE CLASSIFICATION OF SEDIMENTARY
MATERIALS*

Name	Grade limits (*diameter in mm*)
Boulder	>256
Cobble	256–64
Pebble	64–4
Granule	4–2
Very coarse sand	2–1
Coarse sand	1–1/2
Medium sand	1/2–1/4
Fine sand	1/4–1/8
Very fine sand	1/8–1/16
Silt	1/16–1/256
Clay	<1/256

Mud—a term used for a mixture of silt and clay-sized particles with varying amounts of sand and organic debris

---

* From W.C. Krumbein and F.J. Pettijohn, *Manual of Sedimentary Petrography* (New York: Appleton-Century-Crofts, 1938), p. 80.

TABLE 14-2

SETTLING OF SEDIMENTARY MATERIALS
(After Gilluly, Waters, and Woodford; 1957)

Diameter	Rate of settling in sea water ($S = 34\%_0$, $T = 10° C$)	Distance traveled in sinking 1,000 meters through a current of 1 cm/sec
Very fine sand (0.1 mm)	1472 cm/hr	2.4 km
Silt (0.05 mm)	31	116
Clay (0.001 mm)	0.147	24,500

During their transportation in streams, rivers, and in the ocean, terrestrial components tend to be graded according to size and density. The sizes of sedimentary materials are classified as in Table 14-1. Because large particles are more difficult to pick up and transport than fine particles, they settle out of a water column sooner. Table 14-2 shows the effect of size on settling rate. Similarly, dense mineral grains settle sooner than less dense grains, such as quartz. The concept of grading is a fundamental one in marine geology because it relates the distribution of grain size, density, and composition of terrigenous components to the processes of transportation and sedimentation that have influenced the formation of a sediment.

The principle of density grading likewise applies to wind-carried terrigenous components. The wind is less effective as a carrying agent, however, so only the finest materials (silt and clay-sized materials) can be carried for an appreciable distance. The net effect is that considerable amounts of small materials are carried quite far by high-altitude winds. Windborne quartz fragments that occur in deep ocean sediments are concentrated at 30 degrees latitude. This zone contains most of the world's deserts, areas where wind erosion and wind transport are quite pronounced. It is significant that the amount of quartz in sediments decreases below the upper layers. We can thus conclude that deserts were not as widespread in the past as they are today and that climate in the Tertiary period was more humid than at present.

Into the sea along a glaciated coast, glaciers dump undifferentiated material of all sizes from clay to boulders. These terrigenous components, resulting from the action of glaciers, are found in patches close to where the glaciers enter the ocean. Icebergs, however, can cause *rafting* of all sizes of material, because they break off from glaciers and can drift a long way. By this means, *erratic* glacial material can be part of the terrestrial component of a marine sediment that is formed at a considerable distance from a continent (Fig. 14-2).

FIGURE 14-2
Some rocks on the sea floor are distributed in a central cluster of large rocks surrounded by smaller ones.   The pattern is probably the final disposition of a load of glacial debris dumped by a mass of ice.
(Photograph courtesy of U.S. Naval Research Laboratory.)

BIOLOGIC COMPONENTS

The other major sediment-forming component is of biological or organic origin. Some plants and animals living in the surface waters of the ocean have skeletons that sink to the bottom when the organism dies or reproduces. These skeletons become part of the marine sediment. The most important organisms following this pattern are (1) globigerinids, microscopic animals having a shell or *test* of calcium carbonate; (2) diatoms, plants having skeletons or frustules of silica; (3) radiolarians, animals having a test of silica; (4) pteropods, animals having a calcareous shell; (5) coccolithophores, algae that secrete a skeleton made of tiny plates of calcium carbonate; and (6) the extinct *Discoasters*. Examples of these organisms are illustrated on Plates II and III in Chapter 11.

Other organisms contribute only very small quantities to the biological component of marine sediment. Protoplasm from marine organisms sometimes contributes 1 to 2 per cent of the material in marine sediment. Marine algae, such as *Porolithon* and *Halimeda* (Fig. 14-3), secrete skeletal structures of calcium carbonate, and they contribute considerable amounts of biological material in certain tropical regions of the world ocean. On a local basis, shell material from pelecypods and gastropods is an important biologic component.

Biological components of marine sediments are introduced into the ocean wherever sediment-producing organisms live. The distribution of these sediments depends upon (1) the life requirements of these organisms, (2) the chemistry of their skeletons and (3) the depth of the water into which the remains settle.

The life requirements of an organism include, among other things, the chemical and physical properties of seawater. Because these properties are not distributed uniformly in the world ocean, particular organisms flourish in certain areas and are scarce in others.

In a similar way, water depth and body chemistry affect distribution of biological components. Generally, sediments containing calcareous

251

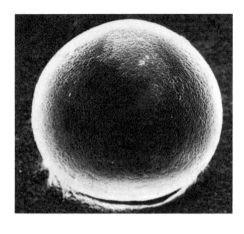

material are not found below a depth of about 4,800 meters. Siliceous
material disappears at a slightly greater depth. Apparently, the temperature, pressure, and CO_2 content of bottom water and the activity of bottom-dwelling organisms lead to the solution of these materials. It is also
generally observed that individual marine sediment samples rarely are rich
in both siliceous and calcareous biologic components. In most cases, the
reason lies in the geographical restriction of siliceous organisms.

EXTRATERRESTRIAL COMPONENTS

It is estimated that each day 10 thousand to 100 thousand tons of
meteorites and cosmic dust fall on the surface of the earth. The cosmic
material that falls into the ocean is often identifiable as magnetic spherules.
Figure 14-4 shows a photomicrograph of such spherules. In addition,
the nickel content of some sediments suggests that they are made
largely of cosmic dust. In some regions extraterrestrial materials form an
important component of marine sediments. Nevertheless, because of the
way in which extraterrestrial components are introduced into the ocean,
they can occur on any part of the ocean bottom.

SUBMARINE VOLCANIC COMPONENTS

Volcanoes have been observed erupting from the sea floor off Japan,
off the western coast of Europe, and off the coast of Iceland (Fig. 14-5).
Near volcanic islands, volcanic rock debris that ranges from ash to boulders
of considerable size are added to marine sediment layers. The ash is often
grayish white to tan and is usually glassy when fresh. Larger rock fragments are usually dark gray to black (basaltic) and lack a vitreous or
glassy appearance.

Since volcanic components enter the world ocean after local eruptions, these components are found in abundance only in submarine volcanic provinces (Fig. 2-16). In the South China Sea, ash from the 1883
explosion of Krakatoa is recognized even today. Ash from the 1912
Katmai eruption is used to correlate marine sediment strata in the Gulf

252

of Alaska. Two eruptions of the Santorini volcano north of Crete in the Aegean Sea can be traced in sediments of the eastern part of the Mediterranean Sea. The upper ash layer in this area resulted from an eruption less than 5,000 years ago (possibly the origin of the Atlantis legend). The lower ash layer occurred 25,000 years ago and has a wider distribution eastward toward Sicily. Often ash deposits become altered by submarine weathering so that volcanic components are recognized by the presence of an assemblage of clay minerals (montmorillonite), phillipsite, and palagonite, a yellowish-brown to deep brown material consisting of a mixture of fresh and altered basaltic glass (Fig. 14-6).

CHEMICAL COMPONENTS

Under some circumstances, dissolved rock material brought into the sea will not remain dissolved. Eventually, it precipitates as a solid and forms a chemical component of marine sediment. The metal-rich manganese nodules (shown in Fig. 14-7A) and the minerals pyrite, dolomite, and aragonite are examples. The mineral phillipsite, previously mentioned as a chemical alteration product of volcanic glass, is also included in this category. When minerals are chemically formed in the place where they

FIGURE 14-5
A submarine volcanic eruption forms a new island called Surtsey near Iceland.
(Photograph courtesy of Icelandic Airlines).

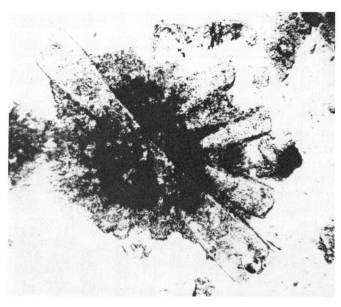

FIGURE 14-6

Microscope-thin section showing basaltic glass with needles of the zeolite, phillipsite, radiating from it. It is believed that phillipsite and montmorillonite are formed by the devitrification of hydrated basaltic glass. Magnification × 190. (Courtesy E. Bonatti, University of Miami).

A

B

FIGURE 14-7A

A section through a manganese nodule dredged from the Pacific Ocean floor showing a core or nucleus of volcanic rock. Other materials may act as nuclei of manganese nodules including shark teeth, and ear bones of whales. Scale equals 1 centimeter.

FIGURE 14-7B

Large, spherical, closely packed nodules covering the seafloor southeast of New Zealand. These nodules are 8-10 cm in diameter. (Photograph courtesy of Lamont-Doherty Geological Observatory, Columbia University.)

are found, the process is called mineral *authigenesis* and the components are *authigenic*. *Allogenic* components are formed elsewhere and transported to the site where they are deposited.

There is a third kind of component—called *diagenetic*. *Diagenesis* occurs when certain substances in the sediment react chemically and are transformed after they become buried beneath later deposits of sediment. In almost every sense, these altered components are metamorphosed—as are metamorphic rocks. Some clay minerals, for instance, undergo diagenetic changes in form and composition in the ocean. Calcium carbonate that has originated in organic life changes its crystal form in response to changes in the temperature and pressure in sediments on the ocean floor. Siliceous remains of marine organisms alter to form *chert*, a dense amorphous variety of silica. Volcanic debris becomes palagonite, phillipsite, and clay minerals. Animal and plant organisms are also responsible directly or indirectly for producing such diagenetic components as glauconite (a greenish mica-like mineral), barite (barium sulfate), and phosphorite. Phosphorite deposits occur frequently in areas of high biological productivity associated with upwelling. Apparently, large amounts of decaying protoplasm on the sea floor favor the formation of this mineral. The processes causing diagenesis in marine sediments are subtle and complex; consequently, they are imperfectly known. It is possible that, through further study of diagenetic effects, scientists will expand the list of diagenetic components to include material now thought to be exclusively allogenic.

Because they are mixed by ocean currents and turbulence, dissolved solids are distributed uniformly in the world ocean. Theoretically, chemical components could occur on any part of the ocean bottom. However, a low rate of deposition of the other components seems to be prerequisite for chemical components to be present in abundance.

In particular local areas, the bottom may be covered with a single chemical component. Manganese dioxide, for example, forms nodules, grains, slabs, coatings, and impregnations on many areas of the ocean floor. This material may be more prevalent than previously thought, because more and more of it is discovered as the exploration of the ocean bottom continues. Manganese (Mn) nodules are most frequently found at the sediment-water interface, although some do exist at depth in marine sediments.

Scientists do not know how Mn nodules are formed. It is apparent that dissolved manganese becomes associated with terrestrial and biological sediment-forming components and is carried to the bottom with them. Another possible source of the manganese could be volcanic debris buried by sediments. After burial, the manganese apparently migrates upward through the sediment column via passages or *interstices* filled with trapped (*interstitial*) water. At or near the sediment-water interface, the manganese precipitates and forms a coating on any available hard surface.

Photographs of the sea floor (Fig. 14-7B) show that manganese nodules as large as 25 cm in diameter litter the ocean bottom in some areas. These accumulations can be valuable economically, because they contain not only manganese but also appreciable quantities of iron, copper, cobalt, and nickel and traces of lead, zinc, and molybdenum. Thus, the nodules represent enormous tonnages of potential low-grade ore awaiting only the development of suitable marine mining technology for their recovery.

Another mineral, pyrite, is found precipitated in sediments that exist in a reducing environment, such as in restricted basins or in sediments rich in decaying protoplasm. Pyrite has been found deposited within the tests of globigerinds in sediments that have contact with oxygenated sea-water. This curious occurrence demonstrates that chemical reactions in the ocean probably involve elaborate biochemical and geochemical processes. For example, deep in the Red Sea, there are warm brines that contain more metal than does most seawater. When we understand the fate of these dissolved metals as the brines mix with cooler water, we will learn much about the chemistry of manganese and other trace metals in seawater.

14.2   CLASSIFICATION OF MARINE SEDIMENT

The proportions of the various sediment-producing components vary from place to place in the ocean. For classification, one component must be sufficiently abundant in the sediment, which is then named according to that component. Common sediment types observed on the bottom of the ocean are listed in Table 14-3.

Sediments are grouped broadly into *neritic* or *pelagic* types according to whether they form by falling from the water on the continental terrace (neritic) or from the water of the open ocean (pelagic). The relationship between this system of classification and the breakdown on the basis of components is shown in Fig. 14-8.

14.3   THE DISTRIBUTION OF MARINE SEDIMENTS

The distribution of each sediment type is not a haphazard one; there are reasons why the sediments are found where they are. We can deduce much about the nature of marine sediments if we consider how all the components of a particular sediment type enter the ocean and how they are transported until they finally form part of a marine sediment.

Most of the components are deposited close to where they are introduced into the ocean. Sand is deposited quite close to shore, on

TABLE 14-3

MARINE SEDIMENT TYPES

Component	Material	Minimum amount required (*by weight*)	Sediment type	
Terrigenous	Clay	70%	Brown clay	
	Sand	70–80%	Terrigenous sands	
	Silt and clay mixture, some sand	70–80%	Terrigenous muds	
Biological	Globigerinids	30%*	Globigerina ooze ⎫	Calcareous
	Pteropods	30%*	Pteropod ooze ⎬	oozes
	Coccoliths	30%*	Coccolith ooze ⎭	
	Diatoms	30%*	Diatom ooze ⎫	Siliceous
	Radiolarians	30%*	Radiolarian ooze ⎬	oozes
Chemical	Authigenic minerals and compounds		Authigenic sediments	(e.g., Mn nodules)

* Required amounts of carbonate or silicate, derived mainly from remnants of marine organisms, the dominant form determining the name.

FIGURE 14-8

Classification of marine sediments on a component end-member basis.

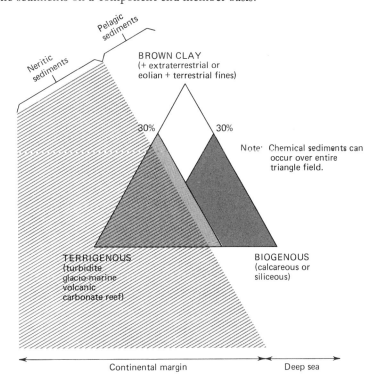

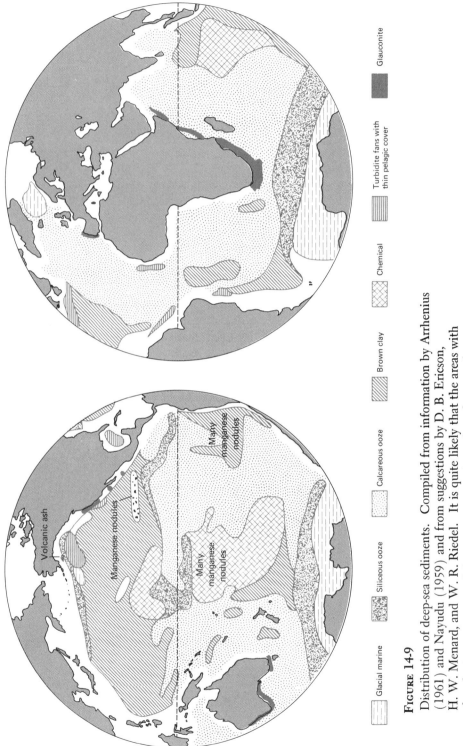

Figure 14-9

Distribution of deep-sea sediments. Compiled from information by Arrhenius (1961) and Nayudu (1959) and from suggestions by D. B. Ericson, H. W. Menard, and W. R. Riedel. It is quite likely that the areas with abundant manganese nodules cover more territory than shown and that there are many more areas where turbidite layers are interbedded with normal pelagic deposits. All boundaries should be considered as subject to extensive changes as information becomes more abundant. (By permission of F. P. Shepard.)

Glacial marine   Siliceous ooze   Calcareous ooze   Brown clay   Chemical   Turbidite fans with thin pelagic cover   Glauconite

beaches, and possibly at the continental margins. Finer material, such as silt and clay, is swept seaward, although silt is deposited closer to shore than clay. The finest clays are able to travel long distances in the ocean and may travel across an ocean basin and back before settling to the bottom. Clay, therefore, is found in all parts of the world ocean. In summary, the amount of terrigenous components decreases seaward away from the continental margins where the bulk of such materials are introduced into the ocean. Larger and heavier particles are concentrated in a band around the continents; finer sediment particles are spread over the entire ocean bottom.

Ocean currents are important in distributing fine terrestrial materials and planktonic, sediment-forming organisms. Throughout the world ocean, turbidity flows apparently cause mixing of several components or reworking of sediments formerly deposited by settling quietly through

FIGURE 14-10

Distribution of sentiment types in a hypothetical ocean basin. A.—profile oriented east-west; B.—profile oriented north-south. Note the importance of depth, latitude and proximity to shore on the distribution of sediment types.

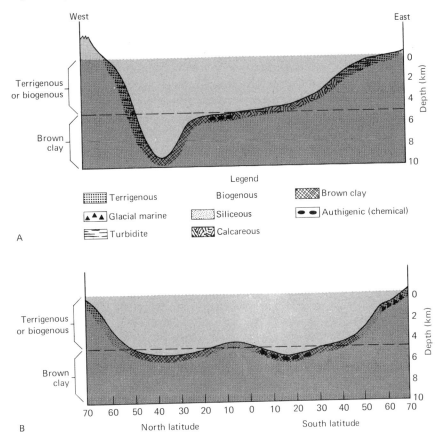

the overlying water. Sediments formed this way, called *turbidites*, are distributed in a way that is predictable despite the fact that the usual distribution patterns of the components in a turbidite are often severely disrupted.

The distribution of sediment types at the bottom of the world ocean is shown in Fig. 14-9. The principles governing this distribution are demonstrated in an ocean basin like the Pacific. An east-west profile across the basin shows sediment types distributed symmetrically as in Fig. 14-10A. At the water's edge, beach sands prevail. Further offshore, terrigenous sands and muds are the dominant sediment type. Coarse terrigenous sediments are deposited close to the continents and fairly well delineate the continental terrace. The continental rise is characterized by turbidite deposits; whereas, beyond the continental margins, biologic sediment types prevail to a depth of about 4.8 kilometers. Brown clay is the dominant sediment type in depths exceeding 4.8 km—that is, the deepest part of the ocean. Brown clay is composed of the finest waterborne terrigenous material and the finest airborne dust. Unusual sediment components, such as extraterrestrial cosmic dust, earbones from whales, teeth of sharks, and clinkers from steamships, are most noticeable in brown clay. Because their rate of deposition is rapid relative to the rate of deposition of brown clay, there is no dilution of exotic sedimentary materials.

The north-south profile in Fig. 14-10B differs from the east-west profile only in the zones occupied by biologic sediment types. Diatomaceous ooze is found in high latitudes because of the life requirements of diatoms. Likewise, radiolarian ooze is largely confined to equatorial regions, because radiolaria abound in warm, low salinity water. The genera of *Globigerina* are found in all latitudes, but individual species have limited geographic distribution. Apparently, the life requirements of individual species are stringent.

14.4 VARIATION IN MARINE SEDIMENTS

The sediment pattern presented in Fig. 14-10 usually prevails throughout the basins of the world ocean. In some places, however, pronounced discrepancies are found: rock outcrops occur at the shelf break; mud flats are found in the intertidal zone; sediment types on the continental shelf have an irregular distribution; and the fine clays in the deep-ocean basins often contain layers of silt and sand carried there by turbidity flows.

Unusual sediments are found in places in the world ocean near sources of special sedimentary materials or where there are sedimentary processes that alter the basic pattern of sediment distribution. Near carbonate reefs, for instance, there are calcareous sediments consisting almost entirely of pulverized remains of reef organisms. Glacial marine sediments are found in high latitudes and at the arctic and antarctic margins (Fig. 14-9). These sediments are characteristically rich in silt and clay-sized crushed rocks produced by glacial action upon continental rock masses. Icebergs then frequently carry coarse material of pebble size and even larger to the site of deposition. (Fig. 14-2).

Volcanic sediments are found immediately adjacent to areas of volcanic activity, such as the western part of the Atlantic Ocean, Indonesia, the Gulf of Alaska, the seamount province off the Pacific northwest coast of the United States, and water adjacent to equatorial America. These sediments are rich in ash and shards of volcanic glass that have erupted from volcanoes on land and under water. Off the west coast of Central America, there are slabs of material, consisting in part of volcanic glass and phillipsite, and coated with manganese oxide. These slabs are thought to represent an ancient volcanic ash layer that has undergone considerable chemical alteration.

Beaches surrounding the world ocean show the greatest variation in unusual sediments. Most of the sand on a beach is brought to the ocean by rivers, which usually flow over tremendously varied terrain. Consequently, adjacent rivers along a coast discharge characteristically different collections of minerals in their sand-sized suspended load. Furthermore, the rocks from which beach sands are derived may be quite different from the rocks along the coast. The beach sediments mirror this difference.

Sediments in estuaries and tidal flats, however, are abnormally fine-grained, although they are in a coastal location. Tidal flats exist where currents or wave action is too weak to cause sorting and removal of fine sediments introduced by rivers. The fine sediments are trapped there, because they are flocculated and form aggregates where fresh water of the river mixes with saline ocean water. The floccules settle more rapidly than the individual particles of which they are comprised. Their accumulation forms a hump of sediment on the bottom of the estuary. In addition, a wedge of saline water moves upstream along the bottom in response to tides and the hydraulic forces associated with river flow (see Chapter 10). The upstream motion of the salt wedge inhibits the seaward transport of flocs so that the estuarine bottom sediment remains as a more or less permanent feature.

Conditions offshore can also cause local variations in beach sediments. An example is the calcite sand of Daytona Beach, Florida. The source of this material is thought to be chemical precipitation of calcium carbonate just off the coast.

SEDIMENT COLOR

The color of marine sediments varies considerably with location, more specifically with the sedimentary environment. Those beach sands at Daytona, Florida, are almost white. Elsewhere, in the tropics and subtropics, carbonate-reef sands have a cream to tan color. The colors reflect differences in form and genesis of the carbonate materials. At Daytona, the carbonate is precipitated chemically, whereas tropical carbonate-reef sands are comminuted fragments of reef algae and other organisms and therefore contain considerable organic impurities. Terrigenous sediments are usually light olive-green to gray-green or gray-blue. The colors are caused by organic matter and reduced iron compounds. The decomposition of organic compounds consumes oxygen in the sediment and keeps iron in the ferrous (reduced) state. Occasionally, organic matter decomposes completely and imparts a black color to the enclosing sediments. The black color is usually lost when the sediment is exposed to air.

Deep-ocean sediments from the bottom to a few tens of centimeters below usually have a different color than the sediments beneath them. Contact with oxygen in the overlying seawater keeps iron in the ferric (oxidized) state, so colors tend toward yellowish-tan or red-brown tints. Where organic matter is scarce, the oxidized zone may extend for tens of meters below the sea floor.

Sediments from anoxic basins are dark greenish-gray or black. These colors are produced by sulfides or sulfur compounds precipitated in the sediments and sometimes by large quantities of organic matter. Sufficiently pure ash layers are cream to buff. Beach sands are often strikingly colored. On the Island of Hawaii, there are green sands made of olivine crystals weathered from lava and black sands made of black lava fragments. Red sands rich in the mineral garnet occur on New England coasts, and brown sands rich in the mineral monazite are mined for rare earth elements in Southeast Asia. Sands composed of heavy accessory minerals are often concentrated by wave action and occur as dark streaks on lighter-colored beaches (see Fig. 10-13B).

EFFECTS OF TRANSPORT IN TURBIDITY FLOWS

Turbidites are formed when the processes of erosion, transportation, and deposition take place under particular conditions of geography and ocean-bottom configuration. Submarine slides, for instance, occur in areas where sediments accumulate, at the heads of submarine canyons, on the continental slope, on the slopes of seamounts and ridges, and on the walls of trenches. These slides dislodge masses of more or less compact sediments and throw them into suspension, forming a turbidity flow. The seaward extent of a turbidity flow is often determined by bathymetric features. Where submarine trenches or seamount chains occur near the sediment source, they act as barriers to the flow, and so the seaward extent of the resulting turbidite deposit is limited. On the other hand, ocean basins frequented by turbidity currents can contain turbidite deposits thousands of kilometers from land.

The irregular distribution of a turbidite sedimentary sequence is demonstrated by differences in the degree of smoothness of the ocean floor. If a basin floor has no turbidites, it usually has quite rough regions of abyssal hills; whereas, areas with turbidites have smoother bathymetry (i.e., abyssal plains). The rock surface underlying abyssal plains is probably rough but is buried under the thick turbidite sedimentary sequence.

LAYERING IN MARINE SEDIMENTS

Many of the sediments of the world ocean lie in layers of different compositions. Samples of brown clay that are several meters thick often show a uniform composition except near the sediment-water interface. However, samples obtained by drilling through the entire sediment sequence reveal that uniformity is not the rule. Instead, there seems to be an alteration of abyssal clays, carbonate oozes, chert (amorphous silicon dioxide), and turbidites.

Turbidites occur in distinct layers of graded terrestrial mud (or reworked sediments) that alternate with layers of pelagic sediments. The sedimentary characteristics of turbidite layers derive largely from their manner of transport and sedimentation. Turbidites originating from near-shore areas carry into deeper water sedimentary materials having shallow-water, terrestrial characteristics. In the turbidity flow, all sizes of sedimentary materials are transported. When suspended materials settle at the site of deposition, they are graded vertically in the order of size and density. The coarse, heavy particles settle first; the fine, light particles, last. Between flows, the pelagic interbeds deposit materials in the same scheme of formation and distribution that governs their occurrence elsewhere in the world ocean. The resulting bottom deposit exhibits an alternation in size, color, mineralogical, and biological composition.

Unlayered sediments form in areas where there is a population of bottom-dwelling organisms that rework the sediment by scavenging bottom sediment for food or that dig burrows into the bottom. Layering cannot occur even if a considerable variety of sedimentary materials is deposited. Consequently, a homogeneous unlayered bottom sediment is formed. Often, however, the uppermost layer of these sediments contains evidence of the activity of these organisms. There may be distinct burrows or chaotic mottling of colors of the sediment. These effects extend downward as much as 1 meter from the water-sediment interface.

If there are no bottom-dwelling organisms, sediments will form layers wherever there are changes in the sedimentation conditions. Certain estuaries, marginal seas, and lakes contain *varved*, or thinly laminated, sediments. These sediments lie in thin layers (a few millimeters thick) that are alternately light and dark, reflecting an alternating or cyclic change in sediments reaching the bottom. The most striking laminae are produced by diatoms having periodic blooms (Fig. 14-11). Light layers rich in diatom tests alternate with darker layers containing fewer diatoms. In such sediments, each light-dark pair of layers usually represents a single cycle of diatom growth. In areas where the cycle is annual, each pair—or varve—represents a single year of deposition; thus, these sediments are

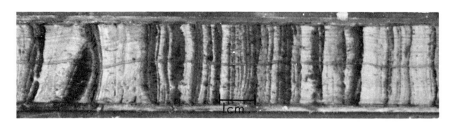

FIGURE 14-11

A cross section of a submarine core showing varves or layers that reflect seasonal differences in deposition in the area. The thickness of the dark layers varies from 1–10 mm. (Photograph courtesy of Joe S. Creager).

easily dated. By dating the sediments and interpreting the oceanic conditions prevailing at the time of the deposition of each sediment layer, scientists learn about the history of the ocean and the earth.

The arrangement of strata, or *stratigraphy*, is studied by examining the layers of ocean sediments; the younger layers, of course, overlie the older ones. The sediment is collected by core samplers that are driven vertically into the bottom (described in Chapter 15). Samples obtained in this fashion usually do not exceed a few meters in length. Much longer ones have been obtained by the deep-sea drilling ship *Glomar Challenger* (see Fig. 15-10). These core samples permit direct observation of the layers of ocean-bottom sediments. In this way, the ocean's history from the Jurassic (160 million years ago) to the present can be studied in detail.

Drill cores from the North Pacific Ocean show that basement rocks are youngest at the west coast of North America where Miocene basalts are found. The basement becomes progressively older toward the west. Jurassic (or older) rocks are found just seaward of the Mariana Trench. The sediments lying on the basement are carbonate (coccolith) oozes. Above these are brown clays. Layers of chert are found at the contact of these two sediment layers and in the carbonate ooze. Clay minerals and zeolites, both from partially altered volcanic ash, are found in layers within the brown-clay layer. The ages of the lowermost sediments parallel those of the basement. Mid-Cretaceous sediments are found in the western North Pacific; Oligocene sediments, in the central North Pacific.

There is little doubt that these sediment samples establish that sea-floor spreading in the North Pacific Ocean has proceeded westward from the East Pacific Rise since at least Jurassic time. Equally dramatic evidence of the sea-floor spreading is found in drill cores obtained in the South Atlantic Ocean. The basement and initial sediments become older the farther away from the Mid-Atlantic Ridge. The ages of these materials indicate a rate of sea-floor spreading nearly identical to that determined by a study of magnetic anomalies in the South Atlantic.

The lithologies of the samples from the South Atlantic are similar to those in the North Pacific; that is, an initial carbonate ooze is overlain first by brown clay then by carbonate sediments. Chert layers are absent in the South Atlantic; but in the North Atlantic and Caribbean, they are present just as in the North Pacific.

In general, the evidence from shorter sediment cores indicates a wider zone of tropical climate during the Tertiary, alternating cold glacial and warmer interglacial stages during the Pleistocene, and a warm condition in the Recent (Holocene) epoch that is not as extensive as in the Tertiary.

This climatic sequence is inferred from differences in the relative concentrations of calcium carbonate in deep-sea cores that were collected from the equatorial Pacific. The deposition of calcium carbonate in the deep sea is a function of both supply by biological productivity near the sea surface and removal by dissolution of the calcium carbonate detritus as it settles to the sea floor. Figure 14-12 shows how the concentration of calcium carbonate in the sediment column is related to relative depth below the sediment-water interface and the latitude. The compensation line refers to the latitude at which dissolution of carbonate detritus just equals its supply.

The sequence of layers exhibiting high (60 per cent) and moderate (1 to 60 per cent) carbonate concentrations are thought to reflect the differences in surface productivity during the Pleistocene epoch. Carbonate-rich layers are related to glacial stages when trade winds were more intense; hence, equatorial upwelling and associated biological productivity were greater. During the interglacial periods, the trade winds died down,

FIGURE 14-12

The east equatorial Pacific Ocean is a site of high biological productivity because of upwelling produced by trade winds. During glacial times the Earth's climatic zones were "compressed," resulting in more intense winds—especially the trade winds. This led to greater upwelling hence greater productivity and more calcium carbonate deposition on the ocean floor. High calcium carbonate concentrations are thus correlated with glacial times in sediments from the east equatorial Pacific. (Modified after Arrhenius, 1963, in *The Sea*, ed. by Hill, by permission of John Wiley & Sons, Inc.)

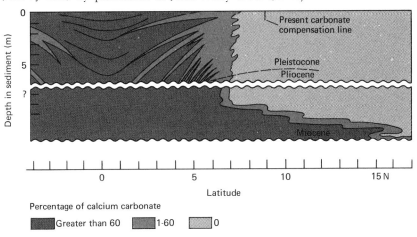

upwelling decreased, and biological productivity slowed. During the Tertiary, the compensation line migrated from 15 degrees N to 7 degrees N latitude. Evidence indicates that this change may reflect a warming trend in the bottom waters (thus decreasing the rate of dissolution) rather than an increase in surface productivity. In the early Tertiary, the compensation line may have extended to 45 degrees latitude. However, the warming trend subsequently decreased, and the compensation line reached its present location by the late Miocene epoch.

There are other indications of climatic control on the nature of pelagic sediments: (1) coiling directions of certain forms of planktonic globigerinids, (2) isotope ratios of oxygen ($0^{16}/0^{18}$), (3) changes in mineralogy and deposition rates of pelagic clays, (4) the distribution of ice-rafted sediment, and (5) the relative abundance of cold-water and warm-water species of planktonic foraminifera, coccoliths, diatoms, radiolarians, or pteropods is another important indicator of past climates. All of these factors must be considered in reconstructing the paleoclimates of the earth.

DATING MARINE SEDIMENTS

The interpretation of marine sediments depends entirely upon ascertaining the age of each of the strata. If there were no method of dating earth materials, scientists could not understand the history of the ocean. However, a system of earth chronology has been developed by dating geological materials. Several techniques are used; each serves to complement the others. The dating techniques most important in oceanography are radionuclide geochronology, paleontological chronology (using fossil remains of plants and animals), magnetic reversal chronology, and chronology of the climatic reversals of the Pleistocene and Recent (Holocene) epochs.

Radionuclide dating is practically the only system that yields age in years, or absolute time. It is based on the phenomenon of radioactive decay, in which a parent radioactive atom emits electromagnetic energy spontaneously and becomes a more stable daughter atom. The rate of decay of parent atoms to daughter atoms at any moment is proportional to the number of parent atoms present at that moment. The proportionality factor is called the *decay rate constant*. This factor is different for each kind of parent atom (or radioactive element), and it determines how long a given amount of radioactive material will last after it begins to decay.

Dating is possible only if the radionuclide is segregated from its daughter atoms at the time of the formation of the material being dated. Once the material has formed, there cannot be migration in or out of it or else erroneous dates are inferred.

Three kinds of radionuclides have been used for dating marine sediments: *primary*, *secondary*, and *induced*. Primary radionuclides decay so slowly that measurable amounts of these materials remain 5 billion years after they formed within our solar system. The daughter products of some primary radionuclides are also radioactive. These secondary radionuclides decay at intermediate rates; that is, measurable amounts remain after hundreds of thousands of years of decay. Induced radionuclides are formed by the bombardment of the earth's atmosphere by cosmic rays. These materials decay rather rapidly but are useful for dating relatively young materials and for tracing ocean currents.

The radionuclides most frequently used to date marine sediments are thorium-230 (Th^{230}), which is also called ionium (Io), protactinium-231 (Pa^{231}), and carbon-14 (C^{14}). Potassium-40 (K^{40}) is also used. The span of geologic ages that can be measured with each radionuclide is shown in Table 14-4.

TABLE 14-4

USEFUL RANGE OF RADIONUCLIDES IN DATING

C^{14}	0 to 40,000 years
Pa^{231}	0 to 150,000 years
Th^{230}	0 to 300,000 years
K^{40}	700,000 to 4.5 billion years

Carbon-14 is useful for geologically young materials (formed since the middle of the last glaciation of the Pleistocene). Additional range into the Pleistocene is obtained with Pa^{231}, Th^{230}, and U^{234}, but the major part of geologic time is measured with the K^{40} radionuclide. The K^{40} technique is applied to igneous rock dating, because the gaseous daughter product argon-40 (Ar^{40}) is driven from molten rock but begins to accumulate in crystal lattices when the rock solidifies and cools below 300° C. The K^{40} technique is not suitable for dating marine sediments, but fortunately it is useful for dating the continental lava flows that reveal the succession of reversals of the earth's magnetic polarity throughout much of geologic time (see Fig. 3-2B). Magnetic reversals are measured in marine sediments and basement lava flows in the world ocean, so a marine sediment that dates back to the mid-Pleistocene or before can be assigned an age expressed in years.

Before the advent of magnetic reversal dating, only *relative* geological ages could be assigned to any (except the youngest) marine sediment. This type of dating was based upon (1) paleontology or climatic changes that were reflected by changes in the relative abundance of clay minerals in sediments, (2) the sequence of seawater temperature reversals indicated by oxygen isotope measurements, or (3) coiling direction of certain foraminifera.

Magnetic reversal dating now permits *absolute* ages to be assigned to marine strata and the oceanographic events they represent. As a result, scientists can understand the history of sedimentation in the ocean and particularly the ages and sequence of events in the latter part of the Pleistocene. Figure 3-3 compares the ages of several important dating horizons used to describe events in the history of the world ocean and permits the correlation of these horizons in marine sediments from different parts of the earth.

14.5   THE THICKNESS OF MARINE SEDIMENTS AND SEDIMENTATION RATES

The determination of the age of sediment layers by the various dating procedures can also help us calculate the rates of accumulation of marine sediments. On the average, sediment layers have a thickness of about 1 kilometer, although it varies from place to place at the bottom of the world ocean. Sediments are about 3 kilometers thick in the Argentine basin and about 50 meters thick on the Mid-Atlantic Ridge. The thickest sediments occur at the continental margin and on the continental terrace. Both the total thickness and sedimentation rate vary according to sediment composition. Brown clay accumulates at considerably less than 1 cm per 1,000 years; globigerina ooze, at about 1 centimeter per 1,000 years; and terrigenous mud, at about 6 centimeters per 1,000 years. At the highest sedimentation rate, the 3 kilometers of sediment in the Argentine basin would accumulate in 50 million years. This is only 1 per cent of geologic time! It is not likely that the sedimentation rate prior to the Pleistocene was far slower than it is today. Apparently, therefore, sediment is being removed. Presumably it is being swept under the continents during sea-floor spreading. If the rate of sea-floor spreading is 5 cm per year (actually quite rapid), then it would take 50 million years for a piece of the floor to travel from its point of origin at the Mid-Atlantic Ridge to the Argentine basin, which is 2,500 km away. In other words, at no time during the history of the South Atlantic Ocean could there be more than a 50-million-years accumulation of sediments on the ocean floor.

NONDEPOSITION

The bottom of the ocean is almost totally covered with layers of more or less fine sediment. Exceptions do occur where strong currents tend to sweep sediments away, such as at the break between the continental shelf and slope. In these cases, rocks are exposed. In other cases, conditions are such that sediments no longer accumulate, so the floor of the ocean is an ancient (or *relict*) deposit formed thousands of years ago.

Continental shelves are outstanding examples of such nondeposition. Sediments in these regions are not graded according to distance from land. Instead, the shelves are covered with patches of coarse terrigeneous materials that show evidence of shallow-water or even subareal and fluvial sedimentary processes. Dating, as well as physical evidence, indicates that these sediments were formed by deposition or even by weathering when the sea level was much lower during Pleistocene glaciation.

Under the normal condition of a relatively wide shelf and moderate influx of sediment from land, insufficient time has elapsed for a recent sedimentary veneer to bury the existing Pleistocene deposits. Hence, many of the sediments of the continental shelf represent a disequilibrium between existing oceanic conditions and the sediment texture. These relict sediments are occasionally redistributed by large storms passing over the continental shelf. Generally, however, their existence is associated with the environmental conditions prevailing during a previous epoch in geological time.

READING LIST

HILL, M.N., ED., *The Sea, Ideas and Observations*, vol. 3 of *The Earth Beneath the Sea*, New York: Interscience Publishers, 1963, 963p.

SHEPARD, F.P., *Submarine Geology, 2nd Edition*, New York: Harper and Row, 1963, 557p.

TRASK, P.D., *Recent Marine Sediments, Revised ed.*, Tulsa: Society of Economic Paleontologists and Mineralogists, Special Publication No. 4, 1955, 736p.

ONE
TWO
THREE
FOUR
FIVE
SIX
SEVEN
EIGHT
NINE
TEN
ELEVEN
TWELVE
THIRTEEN
FOURTEEN
FIFTEEN

oceanographic instruments

The oceanographer has at his disposal a wide variety of instruments that either obtain a sample of seawater or tell something about the oceanic environment—for example, the water depth, or its temperature, or the thickness of the sediment. Many such electronic and mechanical instruments are now being used routinely on board ship. These instruments can collect and record vast amounts of data that are then analyzed by high-speed computers. Oceanographic data, which is collected by scientists throughout the United States, is sorted and stored in national collection centers like the U.S. National Oceanographic Data Center in Washington, D.C. This agency serves as a clearinghouse and makes the data available to all scientists.

15.1 GEOLOGICAL MEASUREMENTS

TOPOGRAPHICAL MEASUREMENTS

As early as 1504, depth measurements appeared on maps. Possibly the first deep-sea sounding was made by Magellan in the Pacific Ocean in 1521. The technique used then, and until the end of the 1800's, was to lower a rope and lead weight to the sea bed and measure the length of

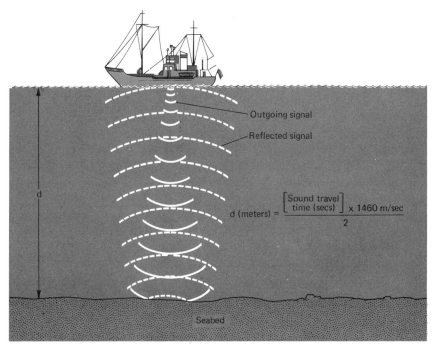

d (meters) = $\dfrac{\left[\begin{array}{c}\text{Sound travel}\\\text{time (secs)}\end{array}\right]}{2}$ × 1460 m/sec

Seabed

Figure 15-1

An echo sounder determines the depth by measuring the time interval
required for a sonic pulse to travel from the ship to the sea floor and back.

Figure 15-2

Two oceanographers study an echogram on board ship.
(Photograph courtesy of Sarah Barnes).

FIGURE 15-3A
An underwater camera and deep-sea housing. (Photograph courtesy of Hydro Products).

FIGURE 15-3B
A lowering frame mounted with stereo cameras ready to lower to the sea floor. (Photograph courtesy of the Office of Information Services, University of Washington.)

rope payed out. In the 1870's, piano wire was used, because it had several advantages over rope: smaller volume, lighter weight, and less drag in the water. Nevertheless, the process of making a deep-sea sounding was difficult and time-consuming.

In the 1920's, an electronic sounding device, called an *echo sounder* or sonic depth finder, was developed. This device measures the time for a sound pulse transmitted from a vessel to travel to the sea bed and return to a listening device. The speed of sound in seawater is known; it is approximately 1,460 meters per second. Thus, the travel time can be converted to a distance indicating the water depth (Fig. 15-1).

The echo sounder was first used as an oceanographic tool by the German research vessel *Meteor* on an expedition from 1927 to 1929. It was better than previous sounding methods, because soundings could be made while the vessel was underway. Hence, a continuous bottom profile (shown in Fig. 15-2) could be obtained. Before this development, there were only several thousand deep-sea soundings. After development of the echo sounder, millions of soundings became available. In fact, it has become standard practice for vessels to make continuous soundings as they sail throughout the world ocean. The compilation of these data has provided the geological oceanographer with an invaluable picture of the surface of the sea floor. The side-scanning *sonar* (Sound Navigation and Ranging) is a further refinement of the echo sounder. It constantly scans the sea floor beneath and to the sides of the observation vessel.

OBSERVATION OF THE SEA FLOOR

Geological oceanographers studying processes of sedimentation in the deep sea routinely use underwater cameras to obtain pictures of the

FIGURE 15-4
Underwater television camera and light, monitor, video tape recorder.
(Photograph courtesy of Hydro Products).

sea floor. Equipment of this sort is also used by biological oceanographers to study marine life. The oceanographer's camera can be lowered to any depth and is prepared to take up to 500 photographs either automatically or upon command (Fig. 15-3A). Cameras are used singly or in pairs that provide stereoscopic viewing of the bottom (Fig. 15-3B).

Closed-circuit television has also been adapted for oceanographic use (Fig. 15-4). A television camera is often used either to moniter the operation of other equipment or to observe phenomena in the marine environment.

Direct observation and sampling by men equipped with scuba gear (self-contained underwater breathing apparatus) is feasible in water no deeper than 45 m (Fig. 15-5). At depths of several kilometers, men in deep-diving submarines are capable of routine observations. The bathyscaphe *Trieste* took men 11 kilometers down into the Mariana Trench.

SEDIMENT SAMPLERS

Several types of sediment samplers are available to the geological oceanographer. He can choose from *dredges, grab samplers,* or *coring*

FIGURE 15-5

A scuba diver exploring the pinnacle of Cobb Seamount 450 km off the Washington Coast. (Photograph courtesy of Walter Sands.)

FIGURE 15-6
A biologist's dredge ready to be
lowered over the side. (Photograph
courtesy of Joe S. Creager).

tubes, depending upon the nature of the bottom in the area of investigation and the degree of sophistication of the sampling program.

A dredge is a strong box-like apparatus that is dragged along the bottom of the sea. It can operate at any depth. A wire-mesh or cloth-mesh lining is placed inside the dredge to keep sedimentary material from being lost. The size of the mesh determines the size of material retained in the dredge (Fig. 15-6). Dredges are also designed to chip rock fragments from submarine rock outcrops.

Grab samplers obtain a relatively unoriented and somewhat disturbed volume of sediment from the layer at the sea floor. Grab samplers also operate at any depth, but they are usually specialized in order to sample a particular sediment type. The Shipek sampler is used if coarse or hard-packed material form the bottom. Other types are used for sampling sand or finer materials (Fig. 15-7). The box sampler (Fig. 15-8) collects large-volume, mildly disturbed samples from the sediment-water interface.

Marine sediment samples for stratigraphic study must be obtained from successive layers extending as deep as practical. These samples must also represent the order of deposition on the bottom. A gravity corer is the simplest instrument that can penetrate marine muds and obtain samples of this kind. It consists of a hollow tube, weighted on top, that is driven into the bottom by the force of gravity (Fig. 15-9A). As it penetrates the bottom, the tube fills with sediments, which remain inside as a core when the tube is brought back to the vessel. Core samplers operate

FIGURE 15-7
A Shipek grab sampler. (Photograph courtesy of Hydro Products).

FIGURE 15-8
A box sampler for collecting undisturbed samples of the sea floor. (Photograph courtesy of G. M. Mfg. and Instrument Corp.).

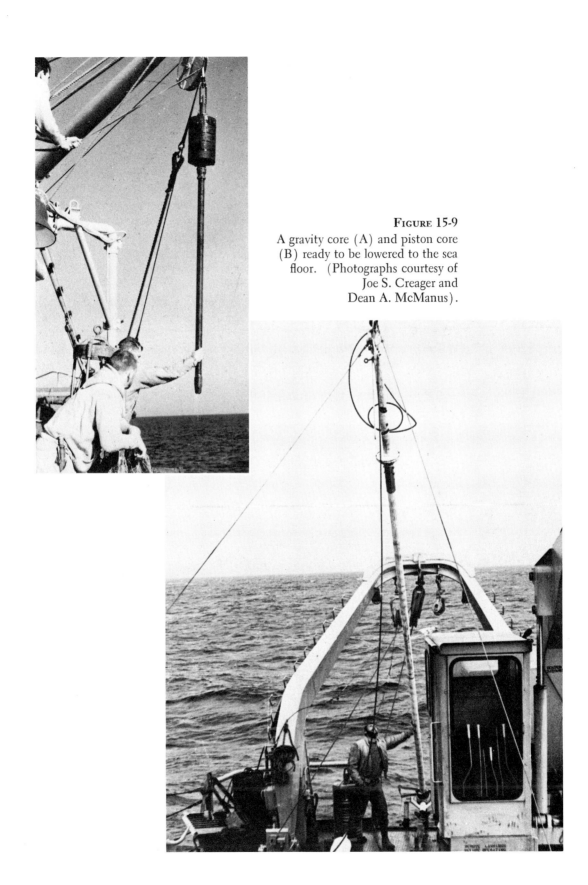

Figure 15-9
A gravity core (A) and piston core (B) ready to be lowered to the sea floor. (Photographs courtesy of Joe S. Creager and Dean A. McManus).

poorly in sandy sediments. However, in silt and clay-sized material, cores as long as 1 m can be obtained regularly.

The piston corer is a more complicated and effective type of coring mechanism. It consists of a coring barrel, weighted on top, with a tightly fitting piston inside. The piston is adjusted on the lowering cable in such a way that, when the barrel reaches the sediment interface, it slides past the piston into the mud (Fig. 15-9B). The piston corer fills with mud more easily than the gravity corer, because a decreased pressure is produced within the barrel of the piston corer. Cores ranging from 25 to 30 m in length are obtained frequently with this device.

Drills, such as percussion drills and various rotary drills, have obtained sediment cores considerably longer than those obtained with gravity or piston corers. The drill ship *Glomar Challenger* has taken core samples as long as 1,000 m from the floor of the open ocean where the water depth exceeds 5 km (Fig. 15-10). This vessel combines a drill similar to a full-scale oil drill with acoustical position-sensing and automatic dynamic-positioning gear. Much of the sedimentary evidence supporting the ideas of the seafloor spreading hypothesis has been obtained by the *Glomar Challenger* during its participation in the Deep-Sea Drilling Project, which was begun in 1969 and is sponsored by the National Science Foundation.

GEOPHYSICAL MEASUREMENTS

Because geophysical measurements have been made on a world-wide basis, we can study the structure of the ocean basins and their relationships

FIGURE 15-10

Deep sea drilling ship *Glomar Challenger*. (Photograph courtesy of Scripps Institution of Oceanography, University of California at San Diego.)

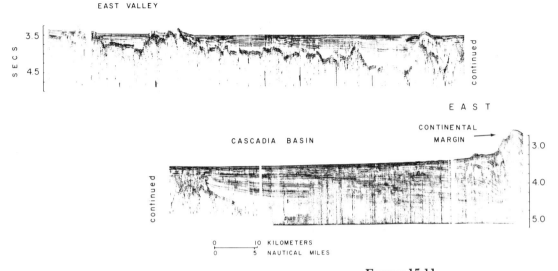

WEST

EAST VALLEY

SECS 3.5 4.5

EAST

CONTINENTAL MARGIN

CASCADIA BASIN

3.0 4.0 5.0

continued

0   10 KILOMETERS
0   5   NAUTICAL MILES

FIGURE 15-11

A typical reflection profile of the sea-bottom in the Northeast
Pacific Ocean Basin. One second of sound penetration represents
approximately 1500 m. (Photograph courtesy of Dean A. McManus).

to the continental masses. In addition, by using these measurements, we
can make a model that accounts for the structure and mechanisms of the
crust of the earth.

A number of different instruments are needed for these measurements. The earth's magnetic field throughout the world ocean is measured by ships carrying magnetometers. Thus, we have a visualization of
the remnant magnetization of the sea floor that gives strong support to the
sea-floor spreading hypotheses. To measure the thermal gradient in the
sediments, sensitive thermal sensors are mounted on sediment core barrels
and thrust into the sea floor. The gradients are interpreted to obtain the
magnitude of heat flow through the sea floor. These data provide information on tectonic processes at work in the world ocean today. Gravity
measurements are made at sea using a pendulum device corrected for the
ship's accelerations. Data on the variation of the magnitude of gravity in
the world ocean reveals how rocks of different density are distributed beneath the seafloor.

Seismic techniques are also used for measurement. For example, the elastic properties of the earth are measured through the propagation of elastic, or seismic, waves that occur naturally as earthquakes or are
produced artificially by high-energy sound. The seismic reflection profiling technique uses sound energy to determine crustal structure. This
technique is similar to echo sounding except that high-energy sound pulses
are transmitted. These pulses not only reflect from the sea floor but also

279

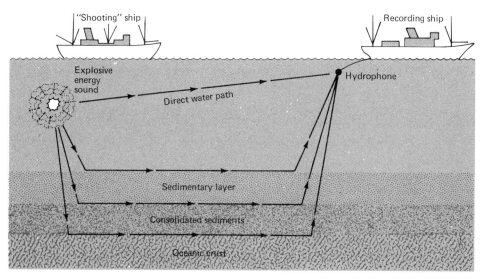

FIGURE 15-12
Seismic refraction technique.

penetrate the bottom. Energy is reflected at discontinuities in the layers of bottom sediments. The result is a profile of all reflecting layers within the range of penetration (Fig. 15-11). Several types of energy sources are used to generate the penetrating signal. Low-frequency sound is not as absorbed by rock as the higher frequencies, but it affords less resolution of small features at the bottom. The signal is produced in the water by an electrical discharge, sudden release of compressed air, an explosion, or any device that converts electrical energy into sonic energy. Each energy source produces a signal of unique acoustical quality, so that its use is limited to particular applications.

Refraction profiling uses the principle that energy impinging upon a sediment layer will refract depending upon the angle of incidence of the energy and its propagation velocity. In a series of strata of different composition, energy propagates rapidly along paths in certain layers and slowly along others (Fig. 15-12). A single pulse of energy refracted by sediment strata becomes separated into a series of pulses that represent the various propagation paths through each layer. In practice one ship produces a seismic pulse and another ship or sensor buoy detects the refracted energy from some distance away. Energy travelling horizontally through the water arrives at the receiving sensor first provided it is close to the seismic source. Energy that penetrates deeply into the bottom may arrive at a sensor deployed several kilometers from the source sooner than the direct signal that passed horizontally through the water. Analyzing the times of arrival of energy at sensors deployed at several distances permits interpreting the geologic structure beneath the sea floor (Fig. 15-13). A refraction profile is constructed by making a series of refraction measurements along a cruise track.

280

15.2   CHEMICAL MEASUREMENTS

Routine measurements of the properties of seawater include temperature, oxygen, salinity, and nutrients (phosphate, nitrate and silicate ions). Techniques for measuring these properties are either semiautomatic or fully automatic; in many cases, the analytical data is reduced by digital computer for storage, collation, and interpretation.

An instrument of great help to chemical and physical oceanographers is the STD (salinity, temperature, and depth) probe, shown in Fig. 15-14. This device consists of a sensing unit, which is lowered through the water column, and a data-receiving unit on shipboard. The sensing unit, containing an electronic thermometer, a pressure transducer, and an induction salinometer, transmits continuous electronic signals through a cable connecting it with the ship. As the sensors are lowered through the water column, the receiving unit plots the signals on graph paper. Thus, the oceanographer obtains an immediate graph of temperature, salinity, and depth while on station. Often, this information is fed directly into a shipboard computer for additional computations and display.

COLLECTION OF SEAWATER SAMPLES

For most chemical determinations, it is necessary to retrieve a sample of seawater from several predetermined depths. The most common sampling device is a Nansen bottle (Fig. 15-15), which is a metal tube (lined with a chemically inert plastic) having valves at both ends. The

FIGURE 15-13
An example of a seismic refraction record.

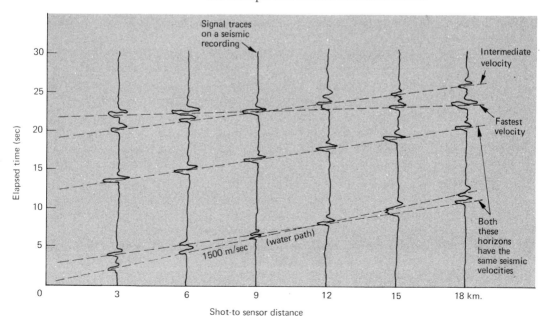

A

B

FIGURE 15-14

An S.T.D. sensing head ready to
be lowered through the water
column (A), while a shipboard
computer concurrently plots
vertical profiles of temperature
and salinity (B).

bottle, with the valves open, is lowered on a wire to the desired depth.
Then a messenger (a metallic weight) is slid down the wire to hit the
bottle, causing it to reverse in situ and close its valves on water. In practice, several bottles (up to 12) are mounted at standard intervals on the
hydrographic wire. When tripped by a messenger, each bottle drops another messenger to the bottle below, thus closing all bottles on the wire in
succession. On board the exploration vessel, seawater samples are transferred to smaller sample bottles and subjected to a variety of chemical
analyses.

TEMPERATURE MEASUREMENTS

The reversing thermometer is the device used most frequently to
measure the temperature of seawater. It is a mercury-in-glass thermometer
that is lowered to the desired depth and inverted (Fig. 15-16). This reversal causes the mercury column to break in such a way that a quantity of

282

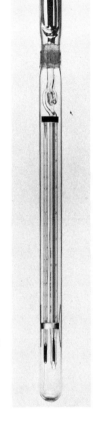

FIGURE 15-15
Nansen bottle mounted on a wire.
The thermometer rack is designed
to hold four thermometers.
(Photograph courtesy of G. M.
Mfg. and Instrument Corp).

FIGURE 15-16
Reversing thermometer in the
reversed position. (Photograph
courtesy of G. M. Mfg. and
Instrument Corp).

mercury representing the in situ temperature is isolated from the mercury reservoir. In practice, the thermometers are mounted on the outside of a Nansen bottle so that the reversing process that closes the valves of the bottle also causes the in situ temperature to be recorded.

Reversing thermometers are manufactured to either respond to or ignore pressure. When one of each type of thermometer is mounted on a Nansen bottle, the difference in the temperature readings indicates the pressure affecting the unprotected thermometer. This pressure is used to calculate the actual depth where the Nansen bottle was reversed. Reversing thermometers are considered reliable only to the nearest 0.02 degree (C).

Nowadays, extremely precise electronic temperature-measuring devices are used routinely. For example, the crystal thermometer utilizes the vibrational frequency of a quartz crystal to indicate temperature. This thermometer can resolve temperature changes as slight as 0.0001 degrees

(C) and has been used to measure the small temperature fluctuations that occur at the deep-sea floor.

SALINITY MEASUREMENTS

Seawater conducts an electrical current according to the temperature, salinity (ion content), and pressure. Therefore, if the temperature and pressure are known, the conductivity of seawater can be used to determine its salinity. Many oceanographic vessels have a *salinity bridge* aboard as standard equipment. This device compares the conductivity of a seawater sample to that of a known, standard sample (Fig. 15-17). If the temperature is carefully controlled, the salinity of the sample is easily determined. The salinity bridge, like other salinometers, must be calibrated with seawater of known salinity. The titration described in Chapter 4 still serves as the fundamental method for determining calibration salinities.

MEASUREMENT OF NUTRIENTS

Biological consumption of nutrients continues in a water sample after it has been collected, so it is necessary either to analyze nutrient ions in a water sample immediately or to arrest changes in nutrient ion concentrations during storage. Nutrient ion determinations chosen for use aboard a ship are simple and can be performed quickly, so that chemists can analyze samples as fast as they are collected.

FIGURE 15-17
Portable salinometer. (Photograph courtesy of G. M. Mfg. and Instrument Corp.).

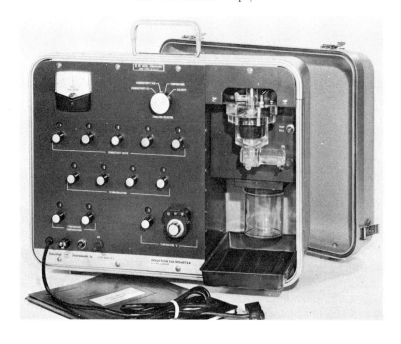

Colorimetry is the usual technique used to measure the concentration of nutrient ions in seawater. The seawater sample is treated chemically in such a way that a color is produced having an intensity depending on the concentration of the nutrient ion. The solution is transferred to a special glass cell and placed in a *spectrophotometer*. In this device, a beam of light passes through the flask, and the relative absorption is measured. Calibration curves are used to convert relative absorption values to nutrient ion concentration. Colorimetric techniques requiring approximately 20 minutes have been developed for determining phosphate, nitrate, and silicate ions in seawater.

However, there are instruments now available that make these determinations virtually automatic. The autoanalyzer shown in Fig. 15-18 samples the seawater collection bottles in sequence, mixes reagents, and measures relative light absorption. This instrument can perform as many as 60 samplings and analyses per hour. The measured concentrations of each sample are plotted on a graphic recorder, or the results can be fed directly into a computer.

15.3   PHYSICAL MEASUREMENTS

MEASUREMENT OF OCEAN CURRENTS

Oceanographers use a variety of methods for measuring ocean currents. Some current-measuring techniques are simple. But making detailed measurements of currents in the open ocean is difficult. For example, let us assume that we want to obtain a one-month record of the South Equatorial Current in the middle of the Pacific Ocean. First, we must have devices that will operate and record continuously for that period. The parts of the meters must be corrosion-resistant, strong, and reliable. Secondly, we must anchor the meters, so they remain fixed with respect to the sea floor and orient themselves properly with respect to the current flow despite severe attack by storm waves. Furthermore, the instruments must be retrievable. We will have to identify the instruments by a radio beacon or flashing light and then collect all equipment. Even this simplified list of considerations indicates that making direct current measurements anywhere in the world ocean requires considerable effort, an advanced marine technology, and sophisticated equipment.

Early indications of the speed and direction of ocean currents were obtained by devices cast adrift at sea. In the early 1900's, scientists usually used drift bottles, which were small bottles that contained an identification card giving the time and place of release and the bottle number. The card requested that the finder fill in the time and place at which the bottle was found and return it to the oceanographic laboratory. A variety of other drifting devices, ships, plastic cards, drift poles and even inflated plastic mattresses have been used to observe currents. Drifting objects have also been used for measuring deep currents. The Swallow float, for instance, is a device that can be weighted to sink to a predetermined depth when thrown overboard. Then it emits sonic signals that allow it to be traced by a vessel as it moves with the deep currents. There are also drifters designed to be bounced and dragged along the bottom by currents at the sea floor.

Another type of mechanical current meter designed for the marine environment is the Ekman meter. This device contains a vane that orients the instrument to the flow of the current and a propeller that rotates as a function of speed (Fig. 15-19). A simple and ingenious mechanism records the rotation rate of the propeller and direction of the meter. A circular container that is free to turn 360 degrees about a vertical axis is located beneath the instrument. This container is a magnetic compass and is partitioned in 10-degree compartments. Above the propeller is a reservoir of small metal balls. The meter is lowered to some depth and is activated by a messenger. The mechanism is so designed that for each 100 turns of the propeller, a ball is dropped from the reservoir into the magnetically oriented compartments. After a given time, another messenger deactivates the propeller, and the instrument is then returned to the ship. The number of balls that dropped during the measurement time indicates the average speed at that depth. The particular compartments into which the balls fell indicate the direction of the flow relative to magnetic north.

The Savonius Rotor current meter measures the rotation rate of an impeller, which has a shape completely different from a propeller. These meters are omnidirectional and hence do not have to be oriented to the current flow. Since they are sensitive at low speeds (approximately 1 centimeter per second), they are used for deep-sea current measurements. To measure flow direction, a separate vane must be placed in the vicinity of the rotor. In Fig. 15-20, the rotor and direction vane are coupled to a shipboard unit that gives a continual visual display of the speed and direction of the current. This complete unit is light enough to lower by hand from a small boat. The deck unit can be coupled directly to a recording device if desired.

MEASUREMENT OF WAVES AND TIDES

When a wave passes through shallow water, it is associated with a fluctuation of hydrostatic pressure at the bottom. The pressure increases under the wave crest and decreases under the trough. The magnitude of the pressure change is a function of the wave characteristics and the depth of water. This effect makes it possible to use a pressure-measuring device as a wave gauge. Pressure transducers consist of specially packaged strain gauges or potentiometers in which the pressure exerted on a metal diaphragm causes a change in a calibrated electronic signal. Thus, changes in signal strength or frequency can be recorded as pressure fluctuations. One such device is shown in Fig. 15-21.

15.4 BIOLOGICAL MEASUREMENTS

Biological oceanographers have developed a variety of mechanisms to obtain specimens from all parts of the water column and the sea floor. Representative devices are discussed here according to the mode of life sampled.

COLLECTION OF THE BENTHOS

For collecting benthic organisms, geological and biological oceanographers use similar devices. The benthos is sampled with dredges (Fig. 15-6), which are dragged along the bottom and indiscriminately collect material in their path. Samples of the benthos are also made with grab samplers (Fig. 15-8) that obtain those organisms living wherever the sampler happens to fall. The ocean bottom and the benthic organisms living there can be observed by underwater cameras (Fig. 15-3) and by underwater television devices (Fig. 15-4).

COLLECTION OF THE PLANKTON

The sampling of plankton is done by some form of filtration. A variety of fine-mesh nets, such as the half-meter net (Fig. 15-22) and the

FIGURE 15-18

This automatic analyzer can continuously analyze sea-water samples for nitrate, phosphate, silicate, and ammonium concentrations. (Photograph courtesy of the Bureau of Commercial Fisheries.)

Recorders for:

Nitrate Silicate   Phosphate ammonium

Reagents

Pump

Manifolds

Colorimeters for each nutrient

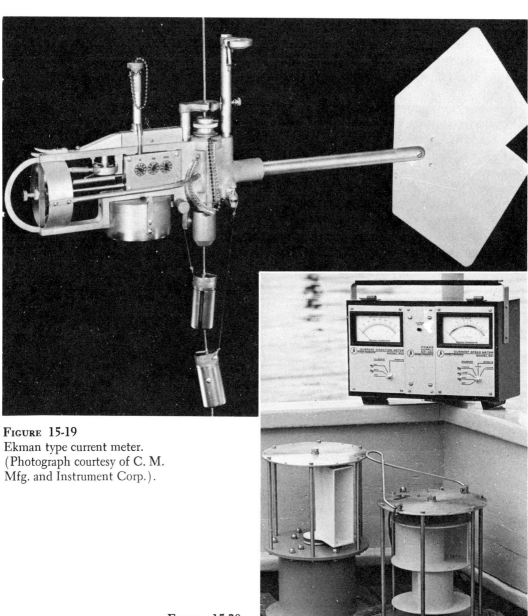

Figure 15-19
Ekman type current meter.
(Photograph courtesy of C. M.
Mfg. and Instrument Corp.).

Figure 15-20
Savonius rotor type current
meter. (Photograph courtesy of
Hydro Products).

Clarke-Bumpus sampler (Fig. 15-23), are lowered to the ocean and dragged behind a moving ship. The plankton pump is another variation. A hose that can be lowered to any desired depth is attached to a pump that draws water from depth and discharges it into a barrel containing a graduated series of nets. With this instrument, the distribution of plankton at depth

288

FIGURE 15-21
In-situ wave and tide recorder.
(Photograph courtesy of Hydro
Products).

FIGURE 15-22
A ½-meter plankton net being
cleaned in preparation of a tow.

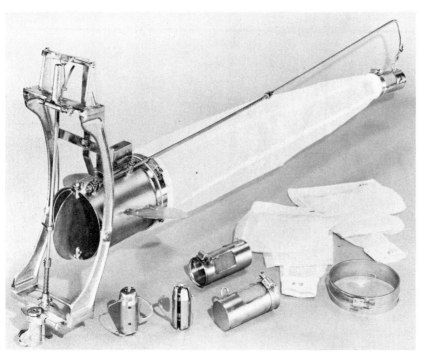

FIGURE 15-23

A Clarke-Bumpus automatic plankton sampler that can be opened and closed at depth by the use of wire messengers. (Photograph courtesy of G. M. Mfg. and Instrument Corp.).

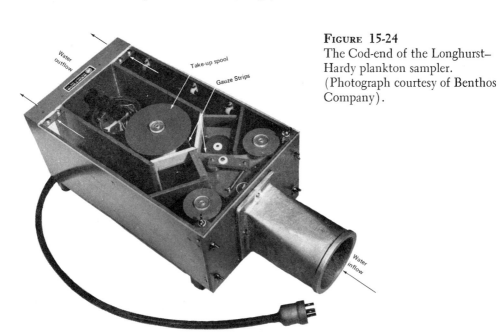

FIGURE 15-24

The Cod-end of the Longhurst–Hardy plankton sampler. (Photograph courtesy of Benthos Company).

FIGURE 15-25
Oceanographic vessel preparing to lower a beam trawl.
(Photograph courtesy of T. S. English.)

and with time can be sampled. The continuous plankton sampler (Fig. 15-24) is a filtering mechanism that is dragged at a predetermined depth behind a moving ship. It filters a sample of water on a continuously moving roll of netting. This apparatus provides a lateral profile of plankton distribution in surface water along a known path of travel over the world ocean. Nannoplankton, however, are too small to be effectively filtered through a net. These organisms are collected in water samples and either centrifuged or pumped through a special filtering apparatus that filters out organisms as small as 0.5 micrometers.

COLLECTION OF THE NEKTON

The nekton are gathered with various trawls or coarse nets. The nets are drawn through the water at high velocity to trap organisms that could otherwise evade capture. In the trawls, there are arrangements of doors or ridged bars to hold the mouth of the trawl open and depressors to prevent the net from inadvertently rising to the surface (Fig. 15-25). Traps (Fig. 15-26) are used in certain cases where organisms can be lured into the apparatus and thereby sampled.

PRODUCTIVITY MEASUREMENTS

In practice, primary productivity is measured by observing the change in any of the substances involved in photosynthesis. In the oxygen method, a water sample containing phytoplankton is obtained and

291

split into two bottles: one is light; the other is dark. The dissolved oxygen in both bottles is measured to verify that both subsamples have the same initial oxygen concentration. Both bottles are returned to the ocean for a period of time. Then the oxygen in both bottles is remeasured. In the light bottle, photosynthesis and respiration show a net productivity, whereas the dark bottle shows the effects of respiration only. The concentration of dissolved oxygen in the water in the light bottle minus that in the dark bottle equals the gross photosynthetic production.

READING LIST

BARNES, H., *Oceanography and Marine Biology, A Book of Techniques.* New York: The Macmillan Company, 1959. 218p.

Instruction Manual for Oceanographic Observations (3rd ed.), U.S. Naval Oceanographic Office Publication 607. Washington, D.C.: Government Printing Office, 1968. 180p.

ISSACS, J.D., AND C.O. ISELIN, *Symposium on Oceanographic Instrumentation,* National Academy of Sciences–National Research Council Publication 309. Washington, D.C., 1952. 233p.

appendix:
physical and
chemical concepts

The reader with no background in the physical sciences, particularly chemistry and physics, may find some parts of the text difficult to understand because of a language barrier. Therefore, in the Appendix, we present explanations of some of the terms and elementary concepts used in several chapters. However, this section is not intended to be a comprehensive treatment, which would be found in introductory textbooks on physics or chemistry.

A.1   SOME INITIAL CONSIDERATIONS

THE METRIC SYSTEM

Most scientists use one of the various metric systems of measurement. Generally, oceanographers use the *centimeter-gram-second*, or cgs, system for measuring length, mass, and time. This system should become familiar to nonscientists because of its consistency in nomenclature (denoting powers of 10) and certain other advantages over the so-called English or U.S. system of measurements.

In the metric system, both the magnitude of any physical property (mass, volume, length, etc.) and the fundamental measure of that property

TABLE A-1

COMMON PREFIXES USED
IN METRIC SYSTEMS

kilo = 1,000 units
deci = 1/10 part of a unit
centi = 1/100 part of a unit
milli = 1/1,000 part of a unit
micro = 1/1,000,000 part of a unit

TABLE A-2

SOME COMMON METRIC LENGTH UNITS (ABBREVIATIONS) AND THEIR
U.S. EQUIVALENTS

kilometer (km)=1,000 meters=3,281 feet=0.62 statute miles =0.54 nautical miles
meter (m)=100 centimeters=3.28 feet=39.4 inches
centimeter (cm)=10 millimeters=0.39 inches=0.00328 feet
millimeter (mm)=1,000 microns=0.039 inches
micron (μ)=1/1,000,000 meter

are named. The root of the name defines the property (length, time, mass, etc.); the prefix denotes the order of magnitude (Table A-1).

LENGTH. The fundamental metric unit of length is the *meter*, defined as 1,650,763.73 wavelengths of Kr^{86} orange-red radiation. The length of the meter is very close to 1/40,000,000 the circumference of the earth. Table A-2 gives the most important metric units of length and some approximate equivalents.

Two units, the *fathom* and the *nautical mile*, persist in oceanography; they are holdovers from early maritime usage. The fathom is equal to 6 ft. The nautical mile is equal to 1,852 meters, approximately 1 minute of arc on any great circle route on the earth's surface These units are still found on most nautical charts printed by the U.S. Coast and Geodetic Survey.

MASS. Mass is most simply defined as a quantity of matter. In the cgs system, the primary quantity is the *gram*, which is equivalent to the mass of 1 cu cm of pure water at 4° C. Other common units of the gram are the kilogram and milligram (see Table A-1).

TIME. The fundamental unit of time, the *second*, is approximately equal to 1/86,400 part of a mean solar day. The other subdivisions of time in the metric system are the same as in the U.S. system. In a geologic sense, time is reckoned in units of thousands, millions, and billions of years. For convenience, geologic time is divided into named intervals, which are shown in Table A-3.

TABLE A-3

GEOLOGIC TIME*

Era	Period	Epoch	Stages	Began Years ago	Dominant life and important events
Cenozoic	Quaternary	Recent (Holocene)		11,000	man
		Pleistocene	Wisconsin	0.2 million	glacial
			Sangamonian	0.4 ″	interglacial
			Illinoisian	0.6 ″	glacial
			Yarmouthian	0.9 ″	interglacial
			Kansan	1.4 ″	glacial
			Aftonian	1.7 ″	interglacial
			Nebraskan	2.0 ″	glacial
	Tertiary	Pliocene		6 million	
		Miocene		26 ″	
		Oligocene		37 ″	mammals
		Eocene		54 ″	
		Paleocene		65 ″	
Mesozoic	Cretaceous			125 million	
	Jurassic			157 ″	reptiles
	Triassic			185 ″	
Paleozoic	Permian			223 million	amphibians
	Pennsylvanian			271 ″	trees
	Mississippian			309 ″	grasses
	Devonian			354 ″	fish
	Silurian			381 ″	
	Ordovician			448 ″	inverte-
	Cambrian			553 ″	brates
Precambrian				1.5 billion	development of oxygen-rich atmosphere
				2+ billion	life originated formation of ocean
				4+ billion	oldest rocks
				5+ billion	origin of earth and solar system

* A person not familiar with the geologic time scale would realize that 5 billion years is a long time but would have no understanding of what that age represents relative to other important occurrences throughout geologic time. Therefore, this table is presented to familiarize the reader with the scope of geologic time and with some important occurrences within this time span.

FORMS OF ENERGY

Physical properties are best understood if they are related to some underlying scheme. The concept of energy is taken as the basic concept in this text, because, by relating physical phenomena this way, we can demonstrate how all physical processes on the earth are ultimately dependent upon the sun.

It seems that the best way to describe energy succinctly is to say that everything in the universe seems to consist of energy. We cannot say what energy is however. Nevertheless, inquiries into the nature of energy have revealed some of its characteristics. One characteristic is that the manifestations of energy occur in several forms. One form is *matter*. Einstein postulated the formula: $E = mc^2$, where c equals the celerity or speed of light (approximately 300,000 kilometers per second), so that any quantity of matter, m, can be represented as a definite amount of energy, E. The validity of this equation was demonstrated dramatically by the explosion of the first atomic bomb in New Mexico on July 16, 1945.

Another manifestation of energy is the form of *radiant energy*. We can show that any radiation—light, for example—is equivalent to a definite quantity of energy by the equation: $E = hf$, where h is Planck's constant and f is the frequency of the radiation (i.e., a particular color of light).

There are some additional types of energy. *Magnetism* and *electricity* are electrical forms. *Chemical energy* is the energy that motivates chemical reactions between chemical substances. *Thermal energy* is energy that we recognize as heat. *Mechanical energy* is energy that matter possesses by virtue of its position in the universe (*potential energy*) or its motion through space (*kinetic energy*). Waves in the ocean represent a combination of potential and kinetic energy. Sound energy is also a form of mechanical energy.

OTHER CHARACTERISTICS OF ENERGY

Another characteristic of energy is that it can change from one form to another. Many examples exist. A battery contains chemical substances that convert chemical energy into electrical energy. Electrical energy can be made to operate an incandescent lamp that emits light (radiant energy) and heat (thermal energy). Imagine yourself in an automobile. If it is parked on a hill, it has potential energy with respect to the bottom of the hill. Should the brakes be released, the potential energy is converted into kinetic energy as the automobile moves down the hill. If you bring the automobile under control by applying the brakes suddenly, the skidding of the tires converts kinetic energy into thermal energy. If you were burned by touching the hot tire, an amount of thermal energy would cause a nervous impulse to be sent to the brain in the form of electrical energy. Your burned finger demonstrates that thermal energy can be converted to chemical energy in coagulating the protein in skin.

Another characteristic of energy is that it is conserved; that is, energy is neither created nor destroyed.

Lastly, energy can be considered to exist in discrete amounts called *quanta*. The size of the quanta is small and depends upon the form of energy considered. Note that just as matter has a particular nature—that is, atoms are made from subatomic particles—so also does energy have a particulate nature.

A.2 PROPERTIES OF ENERGY IN THE FORM OF MATTER

DENSITY

The quantity of matter in a body is termed its *mass*, and the amount of space that the body occupies is its *volume*. Mass is assigned the basic unit of grams; volume is assigned the basic unit of cubic centimeters. These two properties are related and can be considered simultaneously by using the term *density*. The amount of matter in a specified amount of space is the density of the matter under consideration: density (ρ) = mass/volume. The standard density is that of water at a temperature of about 4° C. This density is arbitrarily set as equal to 1 gm per cu cm. The same standard defines mass. One gram of mass is the mass in 1 cu cm of water at 4° C.

The densities of different types of matter differ because their molecules have different masses and different spacing in their structure. The density of a few substances are given in Table A-4.

It is possible to crowd more molecules into a given unit of volume;

TABLE A-4

DENSITY OF SOME COMMON SUBSTANCES

Substance	Density (gm per cu cm)
Pure water at 4° C	1.00
Seawater	approx. 1.03
Gold	19.3
Air	0.0012
Rock (Granite)	2.7
Steel	7.8
Mercury	13.6
Cork	0.2
Ice	0.92

therefore, the density of a single type of matter can vary. Gases, in particular, are readily compressible.

BUOYANCY

The *buoyancy* of an object is its tendency to float in a fluid. According to Archimedes' principle, a body floats if its mass is less than the mass of the water it displaces. Conversely, a body sinks if it contains more mass than the water it displaces.

The volume of water displaced is the same as the volume of the immersed body. An object with a volume of 1 cu cm displaces a volume of water containing approximately 1 gm of mass. The immersed object will float if its mass is less than 1 gm and will sink if its mass is greater than 1 gm. For example, 1 cu cm of iron contains a mass of 7.8 gm. This cube of iron displaces only 1 gm of water, so it will sink. One cubic centimeter of pure ice contains a mass of 0.92 gm; therefore, it floats on water. These phenomena are usually described in terms of density. Bodies more dense than water will sink in it, and bodies less dense than water will float. An object whose density is less than water is said to have *positive buoyancy*, or a tendency to rise toward the surface of the liquid. If an object is more dense than water, it has *negative buoyancy*. *Neutral buoyancy* occurs when the object and the water have the same density. A neutrally buoyant substance will remain where it is placed and has no tendency to rise or sink.

STABILITY

The concepts of buoyancy and density are combined in the concept of *stability*. Stability can be stated as the tendency for a displaced body to return to its original position. The black ball in Fig. A-1A is in a stable position, because it has a tendency to return to its original position if displaced in any direction. An unstable configuration is illustrated in Fig. A-1B. Here, the ball, once displaced, has a tendency to seek a new position rather than return to its original position. Figure A-1C illustrates a configuration of indifferent stability, in which the displaced black ball has no tendency to move in any direction.

In seawater, stability is related to density and therefore to buoyancy. In this case, the black ball in Fig. A-1 is analogous to a small parcel or volume of seawater, and the bowl or plane is analogous to the surrounding water. Figure A-2A represents a stable configuration, because, at every

FIGURE A-1

Diagrammatic representation of the concept of stability.

A. Stable   B. Unstable   C. Indifferent stability

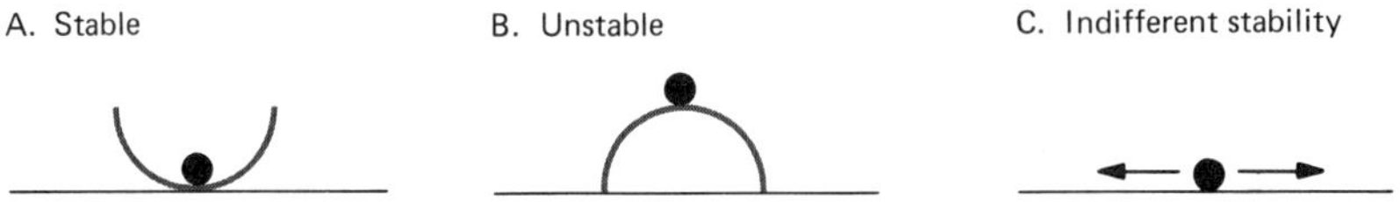

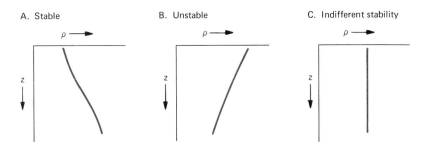

FIGURE A-2

The concept of stability as applied to a water column.

level, water of low density is floating in water of greater density. If a parcel of surface water is displaced downward, there is a tendency for it to rise to the level of its own density—that is, to return to its original position. An unstable configuration is illustrated by Fig. A-2B. Here, water of greater density is floating over less dense water, and there is a tendency for the surface water to sink spontaneously through the low-density water. Likewise, the low-density water tends to rise through the water of higher density. A uniform density profile (Fig. A-2C) is one of indifferent stability. In this situation, a displaced parcel of water has no tendency to return to its original position or to seek a new position.

MECHANICAL ENERGY AND FORCE

Mechanical energy is measured by the work that a body produces. Work is done when matter is moved; the amount of work done equals the distance the body is moved times the force required to make the move. *Force* is commonly thought of as being a push or pull applied to the body. Hence, force is meaningless unless a definite quantity of matter, or mass, is also considered. To understand how force is related to mass, consider a force applied to a stationary object. Initially, the body is at rest and has a velocity equal to zero. One second after applying a certain force to the body, the velocity of the body has increased to, say, 10 cm per sec. In other words, its velocity is changed from 0 cm per sec to 10 cm per sec within the span of a second. This change in velocity is called *acceleration*. If the force on the body is varied, the acceleration varies directly; that is, force is proportional to acceleration for a given object. If the same force is applied to different masses, the acceleration of those masses varies inversely. Mass is proportional to the reciprocal of acceleration; that is, bigger objects accelerate more slowly. If we combine these observations, we obtain the relation between force, mass, and acceleration. This is Newton's second law, often called the equation of motion:

$$F = ma \qquad\qquad [A\text{-}1]$$

So far, we have shown that, in order to measure the amounts of mechanical energy in a body, it is necessary to determine the amount of work being done. To determine work, it is necessary to know the force acting, as calculated from Eq. A-1.

The principle of gravitation states that any two objects in the universe are attracted toward each other by virtue of their mass. The force of attraction is related to the masses of the two objects and to the distance separating them. Newton observed the phenomenon and was able to write the equation.

$$F = \frac{Gmm_e}{r_e^2} \qquad \text{[A-2]}$$

where G equals the universal gravitational constant, m equals the mass of an object on earth, m_e equals the mass of the earth (a constant), and r_e equals the radius of the earth (a constant). The constants in this equation can be combined and designated by g; that is: $m_e \, G/r_e^2 = g$. From this statement is obtained the force of the earth's gravitation upon a mass, m, on the earth's surface:

$$F_G = mg \qquad \text{[A-3]}$$

By comparing this formula with the equation of motion (Eq. A-1), we see that g is an acceleration. Consequently, it is called the acceleration of gravity (on the earth). It has a value of 980 cm per sec per sec.

The principle of gravitation is used in conjunction with the fact that a spring will stretch a distance proportional to the force causing it to stretch. If an object is suspended from a spring, the force of the earth's gravity pulls the object downward. The elongation of the spring is directly proportional to the force that gravity exerts. This force is called the *weight* of the object. Weight is actually a force; that is:

$$\text{weight} = mg \qquad \text{[A-4]}$$

Because g is known on earth, it is possible to calculate the mass of the object from its weight. The unit of force is defined by saying that 1 dyne is the force required to accelerate a 1 gm of mass 1 cm per sec per sec. On the earth, the weight of 1 gm of mass is 980 dynes.

The kinetic energy possessed by a moving body is summarized by the equation:

$$E = \tfrac{1}{2}mv^2 \qquad \text{[A-5]}$$

where v = the velocity of a moving object.

DYNAMIC EQUILIBRIUM

On the earth, gravity also accounts for the acceleration that is assigned to even a stationary object. When all forces acting on a body are in balance—that is, when the net force acting on that body is equal to zero —the body is in *dynamic equilibrium.* Conversely, if the forces are not balanced, its state of motion will change in such a way that dynamic equilibrium is approached. Thus, just because a body is stationary does not mean that no forces are acting on it. Indeed, all bodies on the surface of the earth are attracted to the earth by a force of gravitation; however, the force of gravitation is equally opposed by other forces, so that a net force does not exist. Because the force of gravity is acting on a body at all times, it is always possible to assign an acceleration of gravity to that body.

As another example of dynamic equilibrium, consider an object sinking in water. The force of gravity acting on the object causes a definite acceleration in this case. However, as the velocity increases (as the body accelerates), the friction between the body and the water increases and presents a force opposing the sinking of the object. At a certain velocity, the acceleration by gravity just equals the deceleration by frictional drag, and the body sinks at a constant rate. The net acceleration is zero, implying that the net force acting on the body is also zero, the condition for dynamic equilibrium. It is important to realize that, in this case, the velocity of the body is not zero and that a constant, finite motion exists.

FORCE DIAGRAMS

The forces acting on a body to produce either dynamic equilibrium or disequilibrium are visualized by displaying them graphically in a *force diagram,* sometimes called a *vector diagram.* In force diagrams, forces are depicted as arrows; the orientation of the arrow indicates the direction in which the force acts and the length of the arrow is drawn to be proportional to the magnitude of the force.

In Fig. A-3A, two forces, one twice as large as the other, are shown to be acting in the same direction but on separate bodies. In Fig. A-3B, two equal forces are shown acting in opposite directions on separate bodies.

Where several forces act on a single body, they can be resolved into a single net force. Figure A-3C shows two equal but opposing forces acting to produce a zero net force on a body. In Fig. A-3D, two forces (1 and 2) acting on a body are of unequal magnitude and directions. If a parallelogram (dashed line) is drawn, the resultant (or net) force can be drawn (3) as the combination of forces 1 and 2. This force system can be represented by *either* forces 1 and 2 or by 3 alone, because they are equivalent. Likewise, a simple force can be separated into various component forces. In Fig. A-3E, vector 1 has been divided into a north (3) and a west (2) component force. In this case, force 1 is equivalent to

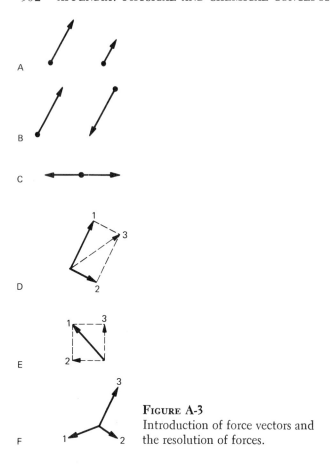

FIGURE A-3
Introduction of force vectors and
the resolution of forces.

forces 2 and 3, and the physical system can be illustrated by either 1 or 2 and 3.

Figure A-3F shows a body acted on by forces 1, 2, and 3. In this example, the net force is zero. This calculation is determined by (1) resolving any two of the three forces and (2) comparing the resultant with the third force. It makes no difference which two forces are picked initially, provided that the resultant force *replaces* the two chosen and is in turn compared to the third force. This technique is important to oceanographers as a tool for analyzing the dynamics of ocean currents. Some actual examples are given in Chapter 7.

PRESSURE

One common force is *pressure*. Strictly speaking, pressure is the force acting on the quantity of matter divided by the area over which that force is acting. A person whose mass is 100 kg (and therefore weighs 98 million dynes) and whose shoes have an area of 300 cm exerts a pressure on the floor of about 32,666 dynes per square centimeter. A unit such as dynes per square centimeter is rather unwieldy; others, therefore, have been devised. The pressure of the earth's atmosphere has been calculated as approximately 1 million dynes per square centimeter. This pressure is called 1 atmosphere. It is called 1 bar by meteorologists and is equal to the pressure at the bottom of a column of mercury 76 cm high, a quantity easily measured with a mercury barometer.

THERMAL ENERGY AND THE STRUCTURE OF MATTER

In matter, the organization of atoms or molecules (henceforth called *particles*) depends upon the phase, or state, in which the matter exists. In the solid state, matter is either crystalline or amorphous. A crystalline material like salt has its particles arranged in space according to a definite geometric pattern. The particles in such a crystal are continuously vibrating. The vibration of any particle occurs around a point in space determined by the position of the particle in the crystal lattice. Amorphous materials like glass differ from crystalline materials in that the particles have no recognizable geometric arrangement. The particles do, however, have vibrational motion around some mean position in space.

In the liquid phase, the particles move farther through space in such a way that any previous position is not "remembered," and the organization of the particles in space is transitory. In addition to vibrating, these particles are free to translate and to rotate around their own axes. In addition, the particles are cohesive; that is, they attract and repel one another in such a manner as to maintain an average spacing between particles and a constant volume. Furthermore, the aggregate of particles in a liquid are influenced by the force of gravity, so that it tends to remain in an open container, such as a cup, and maintain a free surface.

In the gaseous state, particles vibrate and translate too, but they move so far apart that virtually no cohesive attractive forces act between them. For this reason, a gas will not remain in an open container but will tend to leave to fill a larger enclosure. For example, gas that is allowed to escape from a bottle will distribute its particles uniformly throughout the room in which the bottle is situated.

Definitions of these states of matter are merely conventional to some extent. Wax, for example, represents a gradation between a crystalline and an amorphous solid. At certain temperatures and pressures, it is impossible to distinguish between the liquid and gas phases of water. Note, however, that all phases of matter contain mechanical energy manifested as vibrational, rotational, and translational motion of particles. Potential energy is represented by the energy involved in maintaining a mean separation between the particles in matter.

PARTITIONING OF THERMAL ENERGY. When thermal energy is supplied to matter, it is stored in the body of matter in several ways. Part of the thermal energy is used in opposing the attraction of the molecules (potential energy), and part of the thermal energy increases the vibration, rotation, and translation of the molecules. The amount of energy stored in each of these ways depends upon the type of molecule in the body, the phase of the body (gas, liquid, or solid), and the amount of energy already in the body.

TEMPERATURE AND HEAT. The amount of energy that each molecule in a body stores as motion is indicated by the temperature of that body. Heat, on the other hand, is the total amount of thermal energy in a body of matter. We cannot feel how much heat (or thermal energy) a body has any more than we can feel how much potential energy is contained in a rock located at the top of a hill. It is necessary to specify the size of a particular body, its temperature, and its heat capacity in order to state how much thermal energy it contains.

Temperature is the intensity of molecular motion and can be felt as hotness or coldness of a body. The size of a body need not be specified to state its temperature; it is necessary only to bring a thermometer into contact with the body.

THERMOMETERS AND THE MEASUREMENT
OF TEMPERATURE

Two scales are commonly used for the measurement of temperature: Fahrenheit (F) and Celsius (C), also called centigrade. Both of these scales are based on properties of water. Two fixed points, the freezing point and the boiling point, provide a standard for comparison of temperatures. On the Fahrenheit scale, the freezing point of water is 32 degrees and the boiling point is 212 degrees. On the Celsius scale, the freezing point is at 0° and the boiling point, at 100.° It is immediately evident that the size of the Fahrenheit degree is different from that of the Celsius degree. Between the freezing point and boiling point of water, there are 180 Fahrenheit degrees and 100 Celsius degrees. They are in a ratio of 180 to 100 or 9/5. To convert from degrees Celsius to Fahrenheit, multiply the temperature on the Celsius scale by 9/5 and add 32 to account for the fact that the Fahrenheit scale is already at 32 degrees when the Celsius scale is only at 0; that is: $F = 9/5\,C + 32$. To convert from F degrees to C degrees, solve this equation for C: $5/9\,(F - 32) = C$. This relationship is explained in Fig. A-4.

Most temperature-measuring devices are designed to respond to characteristic thermal properties of matter. A change in temperature in a liquid results in a change in the density of that liquid, because the liquid tends to occupy a different volume of space. Mercury has a characteristic, uniform thermal expansion, so it is commonly used in thermometers. Oceanographers, however, measure temperature with electronic thermistors, devices in which electrical resistance is proportional to temperature.

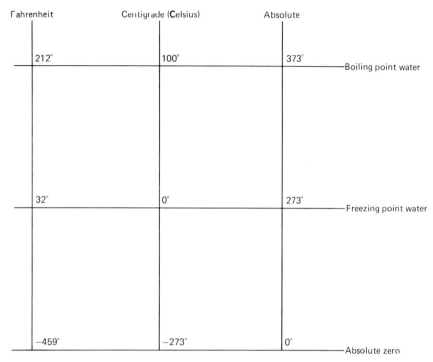

FIGURE A-4

Comparison of different temperature scales.

Scientists sometimes use another scale of temperature, the absolute or Kelvin temperature scale, which is based on the energy of molecular motion. At the zero degree point of the absolute scale, matter has only slight molecular vibration and no translation or rotation. The absolute degree is of the same magnitude as the Celsius degree. The boiling point of water is at 373°absolute, and the freezing point of water is at 273°absolute. Absolute 0°is at −273° C or −469° F.

THE RELATIONSHIP BETWEEN TEMPERATURE AND THERMAL ENERGY

Each kind of molecule stores energy as motion in different amounts; therefore, one unit of heat energy will raise the temperature of one substance more than another. The ratio between the temperature and the amount of thermal energy in a given mass of a substance is called its *heat capacity*. The heat capacity of water is taken as the standard and is arbitrarily assigned the value of 1 degree (C) per gram calorie. Thus, the unit of heat energy is defined as the *calorie*. Note that this calorie is not the one used in the calculation of food energy in dieting; in that case, a Calorie is equal to 1,000 calories. One calorie of heat energy is required

TABLE A-5

HEAT CAPACITY OF SELECTED SUBSTANCES

Substance	Amount of energy absorbed in causing a 1° change in temperature in 1 gm of substance
Ice	0.55 cal/degree C
Mercury	0.03
Copper	0.093
Air	0.33
Rock	0.20

Substance	Number of degrees that temperature of 1 gm of substance can be raised by 1 cal of thermal energy
Water	1°C/cal
Ice	1.8°
Mercury	33.3°
Copper	10.8°
Air	3°
Rock	5°

to raise the temperature of 1 gm of water 1 degree C (in the vicinity of 16°C). The heat capacities of other substances are not as large as that of water. The heat capacities of some substances are listed in Table A-5.

Although water is the standard for defining units of thermal energy, temperature, and heat capacity, it has some of the most unusual thermal properties of all the naturally occurring substances on earth.

LATENT HEAT ENERGY OF FUSION. Part of applied thermal energy opposes attraction of molecules and tends to hold molecules apart. If sufficient heat energy is supplied to the molecules in a solid body, the molecules will stay far enough apart to give liquid properties to it. If still more energy is supplied, the molecules tend to stay even further apart; then the substance becomes a gas. The amount of thermal energy that must be supplied to 1 gm of matter to convert it completely from a solid at its melting point to a liquid at the same temperature is the latent heat energy of fusion. For water, 80 cal are required. It makes no difference which way the phase change proceeds. When water passes from the liquid to the solid phase of water, 80 cal of heat energy per gram are released rather than absorbed.

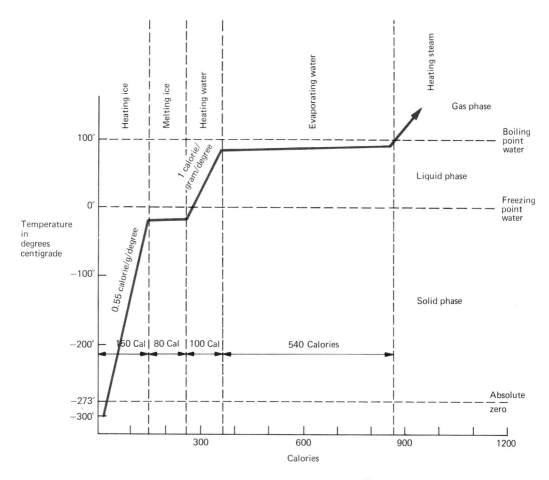

FIGURE A-5

Thermal properties of water. The graph shows the thermal energy (calories) required to raise the temperature of one gram of water from absolute zero to above 100 °C.

LATENT HEAT ENERGY OF EVAPORATION. The amount of thermal energy that must be supplied to 1 gm of matter to convert it completely from a liquid to a gas is the latent heat energy of evaporation. Water requires 540 cal per gm at its boiling point. The latent heat energy of fusion and of evaporation for water is quite high. By comparison, the latent heat energy of evaporation for mercury is 70 cal per gm. The fact that a temperature of phase change is specified does not imply that a phase change must occur at this temperature; water can pass from the liquid to gas phase at any temperature. However, the temperature at phase change determines the exact amount of the latent heat energy absorbed or released at the change of phase. For example, evaporation of water at 0° C requires 595 cal per gm. At 18° C, the phase change absorbs

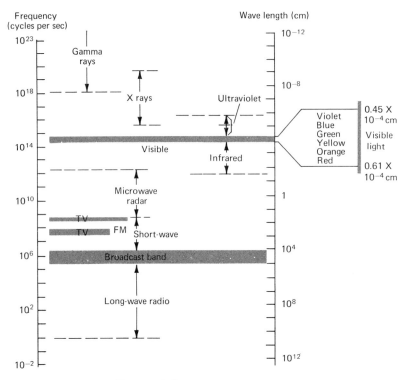

FIGURE A-6

The electromagnetic spectrum.

585 cal per gm. The thermal properties of water are illustrated in Fig. A-5. This curve shows the changes in temperature when heat energy is supplied gradually to 1 gm of water (ice) at absolute zero.

RADIANT ENERGY

Energy in the form of electromagnetic radiation is characterized by either wave or particulate properties. Radiant energy waves can be assigned a constant speed of 300,000 m per second, a frequency in cycles per second, and an amplitude that is a measure of the intensity of the radiation. The same energy can be considered as particles called *photons*. Each photon contains a discrete amount of energy called a *quantum*. Experimental evidence shows that electromagnetic radiation is explained best as waves when observed in one manner and as particles when observed in a different manner. Wave properties of radiant energy are related to particulate properties by the following formula: $E = hf$, where h is 16 times 10^{-35} cal sec and f is the frequency of the radiation. This formula states that the amount of energy in the radiant form is proportional to the frequency of radiation. All possible frequencies of electromagnetic radiation are described in a spectrum, such as shown in Fig. A-6.

Energy is emitted from matter as electromagnetic radiation because

308

of fluctuations in the distribution of energy among the molecules. Col-
lisions of molecules and the vibration of molecules cause these particles
to have first greater, then lesser, energy. Both the *frequency* (energy of
the photon) and the *intensity* (number of photons leaving per second)
are governed by the amount of energy in the matter and, indirectly, by the
temperature of the matter.

Matter will absorb electromagnetic radiations but only in quanta of
certain size (frequency). The size of the absorbed quanta is dependent
upon the molecular structure of the matter absorbing the radiations; that
is, the photon must "fit" into the molecule or it is not absorbed.

A.3   SOME BASIC CHEMICAL DEFINITIONS

CLASSIFICATION OF MATTER

Atoms are the basic particles from which all matter is made. They
have different compositions, depending upon the amount of subatomic
particles contained. Two types of subatomic particles, the *proton* and the
neutron, constitute the nucleus of an atom; the third type, the *electron*,
is outside the nucleus. The number of protons in the nucleus of an atom
determines its chemical nature. The common form of each atom has an
equal number of neutrons and protons in the nucleus. Elements can have
less common forms (*isotopes*), because the nuclei of these atoms may have
more neutrons, hence a greater mass, than the common isotope.

At present, 104 different kinds of atoms are known to exist. Each
kind of atom is called an *element* and is generally considered to be unalter-
able. In truth, alteration can be accomplished by applying a tremendous
amount of energy. Some atoms change by spontaneous radioactive decay;
hence, their elemental properties change with time. Examples of some
common elements are listed in Table A-6.

Two or more atoms of any kind in chemical combination form a

TABLE A-6

SOME COMMON ELEMENTS AND THEIR CHEMICAL SYMBOLS

Aluminum	Al	Iodine	I	Potassium	K
Argon	Ar	Iron	Fe	Protactinium	Pa
Boron	B	Lead	Pb	Silicon	Si
Bromine	Br	Magnesium	Mg	Strontium	Sr
Calcium	Ca	Manganese	Mn	Sulfur	S
Carbon	C	Nitrogen	N	Thorium	Th
Chlorine	Cl	Oxygen	O	Uranium	U
Copper	Cu	Phosphorus	P	Zinc	Zn
Hydrogen	H				

molecule. There are essentially two types of molecules: those formed by the combination of two or more different kinds of atoms, and those formed by the combination of two or more atoms of the same element. Examples of the first type are: water (H_2O), carbon dioxide (CO_2), and sugar ($C_{12}H_{22}O_{11}$). Examples of the second type are: oxygen (O_2), nitrogen (N_2), and hydrogen (H_2).

Matter composed of molecules of the first type is called a *compound,* implying that it is a substance with fixed composition that is separable by chemical means into two or more elements. Compounds are of two general kinds: organic and inorganic. *Organic compounds* contain carbon and can be synthesized by living organisms. They are the building blocks of living matter. Examples are sugars, fats, and proteins. *Inorganic compounds* include all other compounds.

A *mixture* consists of atoms and molecules that are not in chemical combination with each other and have no fixed composition. The ocean illustrates the difference between a compound and a mixture. Water is a compound. Seawater is a solution (a homogeneous mixture) consisting

FIGURE A-7

Relationship between sub-atomic particles, atoms, molecules, solutions, and mixtures. In this scheme, water is a compound, seawater a solution, and the ocean is a mixture.

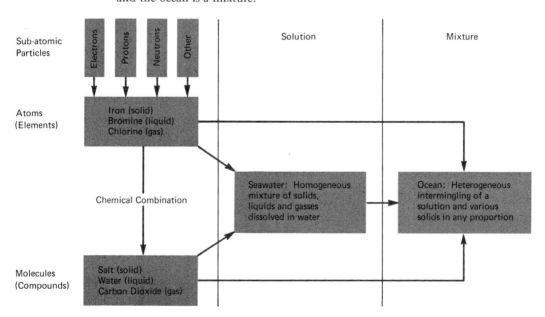

of solids, liquids, and gases (solutes) dissolved in water (the solvent). The ocean is a mixture consisting of a solution and suspended solids. The relationships between these different terms is illustrated in Fig. A-7. *Colloids*, relatively large aggregates of molecules in suspension, are common in the sea; they are intermediate between a solution and a mixture.

A solution has one important property: there is a maximum quantity of solid, liquid, or gaseous material that can be dissolved in it. A solution containing a maximum quantity of solute is said to be *saturated* with that substance. The concentration of this dissolved material is called the *saturation concentration*, or *saturation solubility*.

CHEMICAL NOMENCLATURE

Most people, whether trained in chemistry or not, realize that chemists use symbols rather than the formal names of elements. This chemical shorthand allows a chemist to refer to any form of matter in a simple and concise manner.

The first step in understanding the language of chemistry is to know the symbol for each of the elements. Table A-6 presents the chemical symbols for those elements mentioned in this text.

Next, it is important to learn how these abbreviations for elements fit together to form the names of compounds. Since compounds represent elements in fixed chemical combinations, each different compound has a name and a specific set of symbols to represent the particular combination of elements. For example, a water molecule is composed of two atoms of hydrogen combined chemically with one atom of oxygen. The chemical symbol for water is H_2O. In many cases, the name of the compound refers to the composition, just as its symbol does. Some examples are given in Table A-7.

TABLE A-7

SOME COMMON COMPOUNDS AND THEIR CHEMICAL SYMBOLS

Table salt	NaCl	Calcium carbonate	$CaCO_3$
Carbon dioxide	CO_2	Silicon dioxide	SiO_2
Silver chloride	AgCl	Hydrogen sulfide	H_2S

The concept of an ion is of great importance in chemical oceanography. An *ion* is an atom or molecule that, when put into solution, acquires a net positive or negative charge by gaining or losing one or more electrons. The electron carries a negative charge, so an atom or molecule that has gained one electron will have a net negative charge, and one that has lost an electron will possess a positive charge. The chemical symbol for an ion is made by placing one positive or one negative sign for each electron lost or gained above and to the right of the normal abbreviation for the compound in question. For example, the element sodium readily gives up one electron in solution and becomes the sodium ion (Na^+). Likewise, chlorine in solution gains an electron to become the chloride ion, written Cl^-. Positively charged ions are called *cations*; negatively charged ions are *anions*.

The ions of some metals can exist in solution in two or more states. Iron, for example, can give up two or three electrons in solution, depending upon the demand for electrons imposed by the ions of other elements in the solution. If the demand is low, iron exists in the divalent (2^+) or *reduced* state; if the solution is oxidizing (i.e., has a high electron demand), iron exists in the trivalent (3^+) or *oxidized* state. The iron ions are called *ferrous* and *ferric*; the *-ous* and *-ic* endings are consistently used to indicate the oxidation state of a polyvalent ion.

Acids and *bases* are two important classes of compounds that dissociate and liberate ions in solution. The strength of an acid or base depends upon the completeness of ionization. Acids dissociate into hydrogen (H^+) ions and an anion; bases, into a cation and the hydroxyl (OH^-) ion. The chemical reaction between an acid and a base (the neutralization reaction) produces water and a compound called a *salt*. Most salts ionize readily. The most common salt is sodium chloride (table salt), which ionizes completely in water. There are many such salts, both organic and inorganic. Organic salts generally do not ionize in solution as completely as inorganic salts; indeed, some inorganic salts hardly dissolve at all. A solution containing ions is called an *electrolyte*, which refers to its ability to conduct an electrical current. When a salt dissolves in water, an equal number of positive and negative ions are produced, so that the net charge in the solution is zero. However, the dissolved substance is present as ions and not as salts, even though the term *dissolved salts* is commonly used. Salts can be removed from an ionic solution by freezing or by evaporation, because individual ions do not exist independently in the solid state.

MODES OF CHEMICAL AGGREGATION

Chemical substances come together in solutions, gases, solids, and liquids in several ways. In solutions, there are two general modes of aggregation. One is called a *complex*, in which each chemical substance is capable of independent existence in solution. The other mode includes such discrete chemical entities as ions and dissolved molecules. The degree of association varies in strength and includes (1) strong chemical bonding; (2) hydrogen bonding, wherein pairs of oxygen molecules in water are bonded via a hydrogen atom; (3) formation of ion pairs by electrostatic attraction between the protons (positive charges) of one ion and the electrons (negative charges) of another ion; and (4) chemisorption of ions or charged (polar) aggregates on solid surfaces where there are unsatisfied charges. This last process occurs because atoms at the surface of a substance are not completely surrounded by other atoms of that substance. In the sea, for example, the unsatisfied charges on suspended sediment particles are negative, so cations are usually absorbed on these materials. The bonding is relatively weak. Thus, there can be considerable exchange of cations on these surfaces, particularly where they pass from a river to seawater.

CHEMICALS OF THE EARTH

In nature, discreet chemical aggregations are called *minerals*. A mineral is a naturally occurring inorganic substance with characteristic chemical composition, internal structure, and physical properties. A *rock* is a cohesive aggregate of minerals.

The minerals commonly forming rocks are *quartz*, which is silicon dioxide or silica; *feldspar*, an alkali aluminum silicate; *mica*, a hydrous alumino-silicate; *ferromagnesian minerals*, dark-colored iron and magnesium silicates; and *calcite*, calcium carbonate. Minor amounts of *accessory minerals* are also present in rocks. Accessory minerals have various compositions and are relatively heavy and resistant to decomposition by weathering.

There are several ways in which minerals can be formed into rocks. *Igneous rocks* are mineral aggregates that have solidified upon cooling from a molten state (see Fig. A-8). *Sedimentary rocks* are aggregates of individual mineral or rock particles that have been deposited together and cemented. The sedimentary particles originate from the breakdown of previously existing rocks. Some sedimentary rocks form from minerals chemically precipitated from seawater. *Metamorphic rocks* are mineral aggregates that result when heat, stress, and hot fluids are applied to any preexisting rock—a condition that often occurs after burial of such rocks.

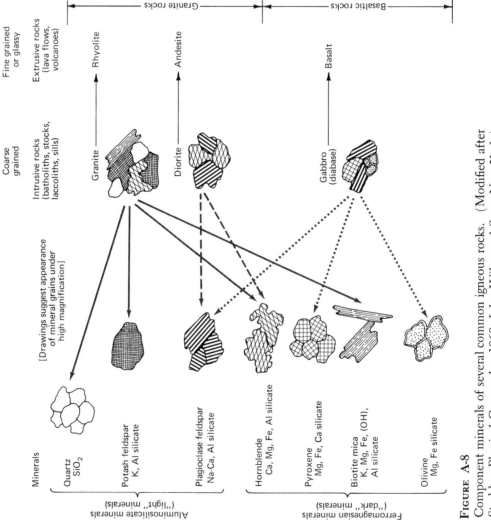

Figure A-8

Component minerals of several common igneous rocks. (Modified after Strahler, Physical Geography, 1960, John Wiley and Sons, New York.)

glossary[*]

[*] In part after B. B. Baker, W. R. Deebel, and R. D. Geisenderfer (eds.), *Glossary of Oceanographic Terms,* 2nd ed., U.S. Naval Oceanographic Office, Washington, D.C. (1966).

glossary

ABYSSAL—(or *abyssobenthic*). Pertaining to the great depths of the ocean, generally below 3,700 meters (2,000 fathoms).

ACCELERATION OF GRAVITY—The acceleration of a freely falling body due to the gravitational attraction of the earth. Its true value varies with latitude, altitude, and the nature of the underlying rocks.

ADVECTION—In oceanography, advection refers to the horizontal or vertical flow of sea water as a current.

AEROBIC—Conditions in which oxygen is present in the environment. Aerobic respiration refers to metabolic processes carried out in the presence of oxygen.

ALGAE—A thallophyte possessing chlorophyll. Algae are single celled, colonial, or multicelled plants having no true roots, leaves or stems.

ALLOGENIC—The term applied to rock or sediment constituents which originated at a different place and at a previous time to the rock of which they now constitute a part.

AMPHIDROMIC POINT—A no-tide or nodal point on a chart of *cotidal lines* from which the cotidal lines radiate.

ANAEROBIC—Conditions in which oxygen is excluded from the environment. Some bacteria can, however, live in these conditions.

ANTARCTIC CONVERGENCE—(or *Antarctic Convergence line, Antarctic Convergence zone*). The Southern Hemisphere *polar convergence*. It is the best defined convergence line in the oceans, being recognized by a relatively rapid northward increase in the surface temperature. It can be traced around the world in the broad belt of open water between Antarctica to the south and Africa, Australia, and South America to the north.

ATOLL—A ring-shaped coral reef that encloses a *lagoon* in which there is no preexisting land, and which is surrounded by the open sea. Low, sandy islands may occur on the reef.

AUTHIGENIC—Formed in the sea. Products of chemical and biochemical action in marine

317

sediments occurring during and after deposition but before burial and consolidation.

AUTOTROPHIC NUTRITION—That process by which an organism manufactures its own food from inorganic compounds.

BALEEN—(or *whalebone*). The horny material growing down from the upper jaw of large plankton-feeding whales, which forms a strainer or filtering organ consisting of numerous plates with fringed edges.

BARRIER BEACH—(also called *offshore barrier*). A *bar* essentially parallel to the *shore*, the crest of which is above *high water*.

BARRIER REEF—A *coral reef* parallel to and separated from the coast by a *lagoon* that is too deep for coral growth.

BASALT—A basic *igneous* (*extrusive*) *rock* composed primarily of calcic plagioclase, pyroxene, and with or without olivine.

BATHYPELAGIC—A depth zone of the ocean which lies between depths of 900 and 3,700 meters (500 and 2,000 fathoms).

BEACH—The zone of unconsolidated material that extends landward from the *low water line* to the place where there is marked change in material or physiographic form . . . . or to the line of permanent vegetation (usually the effective limit of storm waves).

BENTHIC—(also called *benthonic*). 1. That portion of the marine environment inhabited by marine organisms which live permanently in or on the bottom.

2. Pertaining to all submarine bottom terrain regardless of water depth.

BERM—The nearly horizontal portion of a *beach* or *backshore* having an abrupt fall and formed by deposition of material by wave action, that marks the limit of ordinary high tides.

BIOLOGICAL OCEANOGRAPHY—The study of the ocean's plant and animal life in relation to the marine environment, including the effects of habitat, sedimentation, physical and chemical changes in the environment, and other factors on the spatial and temporal distribution of marine organisms, as well as the action of organisms on the environment.

BIOLUMINESCENCE—The production of light by living organisms as a result of a chemical reaction either within certain cells or organs or extracellularly in some form of secretion.

BROWN CLAY—A brownish mud that accumulates in the deep sea and contains less than 30% biogenous remains.

BUFFER SOLUTION—A solution that is resistant to changes in acidity or basicity. A buffer solution contains a weak acid and its salt or a weak base and its salt.

CAPILLARY WAVE—(also called *ripple, capillary ripple*). A wave whose velocity of propagation is controlled primarily by the *surface tension* of the liquid in which the wave is travelling. Water waves of length less than one inch are considered to be capillary waves.

CARBON[14] METHOD—A method of radioactive dating which utilizes the ratio of radiocarbon (Carbon[14]) to Carbon[12] to determine the age of samples containing formerly living matter.

C/G/S SYSTEM—The system of physical measurements in which the fundamental units of length, mass, and time are the centimeter, gram, and second, respectively.

CHALLENGER EXPEDITION—The expedition mounted by the British in *H.M.S. Challenger*, 1873–1876, which made the first extensive oceanographic cruise.

CHART DATUM—(or *datum, tidal datum*). The permanently established surface from which soundings or tide heights are referenced (usually *low water*). The surface is called a tidal datum when referred to a certain phase of tide. In order to provide a factor of safety, some level lower than *mean sea level* is generally selected, such as *mean low water* or *mean lower low water*.

CHEMICAL OCEANOGRAPHY—The study of chemical composition of the dissolved solids and gases, material in suspension, and acidity of ocean waters and their variability both geographically and temporally in relationship to the adjoining domains, namely, the atmosphere and the ocean bottom.

CHERT—A hard siliceous rock of organic or precipitated origin.

CHLORINITY—(symbol Cl). A measure of the chloride content, by mass, of seawater (grams per kilogram of seawater, or parts per mille). Originally chlorinity was defined as the weight of chlorine in grams per kilogram of seawater after the bromides and the iodides had been replaced by chlorides. To make the definition independent of *atomic weight*, chlorinity is now defined as 0.23285233 times the weight of silver equivalent to all the halides.

CHLOROPHYLL—A group of green pigments, identified as *a*, *b*, and *c*, which occur chiefly in bodies called *chloroplasts* and are active in *photosynthesis*. The concentration of each of these pigments has been employed as a means of estimating the rate of photosynthesis (*primary production*).

CLAY—As a size term, refers to sediment particles ranging from 0.0039 to 0.00024 millimeter. Mineralogically, clay is a hydrous aluminum silicate material with plastic properties and a crystal structure. Common clay minerals are *kaolinite*, *montmorillonite*, and *illite*.

COCCOLITH—Very tiny calcareous plates, generally oval and perforated, borne on the surface of some planktonic marine algae (*coccolithophores*).

COMMENSALISM—A symbiotic relationship between two species in which one species is benefitted and the other is not harmed.

CONTINENTAL RISE—A gentle slope with a generally smooth surface, rising toward the foot of the *continental slope*.

CONTINENTAL SHELF—(also called *continental platform*). A zone adjacent to a continent or around an island, and extending from the *low water line* to the depth at which there is usually a marked increase of slope to greater depth.

CONTINENTAL SLOPE—A declivity seaward from a shelf edge into greater depth.

CONVECTION—In general, mass motions within a fluid resulting in transport and mixing of the properties of that fluid. Convection, along with *conduction* and *radiation*, is a principal means of energy transfer.

Distinction is made between: free convection (or gravitational convection), motion caused only by density differences within the fluid, and forced convection, motion induced by mechanical forces such as deflection by a large-scale surface irregularity, turbulent flow caused by friction at the boundary of a fluid, or motion caused by any applied external force.

CONVERGENCE—Situation whereby waters of different origins come together at a point, or more commonly, along a line known as a convergence line. The recognized convergence lines in the oceans are the polar, subtropical, tropical, and equatorial convergence lines. Regions of convergence are also referred to as convergence zones.

CORAL REEF—A ridge or mass of limestone built up of detrital material deposited around a framework of the skeletal remains of *mollusks*, *colonial coral*, and massive *calcareous algae*.

CORIOLIS FORCE—An apparent force on moving particles resulting from the earth's rotation. It causes the moving particles to be deflected to the right of motion in the Northern Hemisphere and to the left in the Southern Hemisphere; the force is proportional to the speed and latitude of the moving particle and cannot change the speed of the particle.

CRUST—The outer shell of the solid earth, the lower limit of which is taken generally to be the *Mohorovičić discontinuity*. The crust varies in thickness from approximately 5 to 7 kilometers under the ocean basins to 35 kilometers under the continents.

DECLINATION—The angle that the sun, moon, planets or stars make with the plane of the equator. The maximum declination of the sun is about 23-½° and of the moon is about 28-½°.

DEEPWATER WAVE—(also called *short wave*). A surface wave the length of which is less than twice the depth of the water.

DENSITY—A property of a substance defined as its mass per unit volume.

DENTRIFICATION—The chemical reduction of nitrate and nitrite compounds to nitrite, ammonia, or free nitrogen. This process is commonly caused by bacteria under anoxic marine conditions.

DESICCATION—Drying up, or loss of moisture.

DETRITUS—(or *debris*). Any loose material produced directly from rock disintegration.

DIAGENESIS—The chemical and physical changes that sediments undergo after their deposition, compaction, cementation, recrystallization, and perhaps replacement, which result in *lithification*.

DIFFUSION—The spreading or scattering of matter under the influence of a concentration gradient with movement from the stronger to the weaker solution.

DIPOLE—An object that is oppositely charged at two points. It usually refers to the charge distribution on a water molecule.

DIURNAL—(*daily*). Actions that recur every twenty-four hours. When considering lunar tides, tides that recur once each lunar day.

DIVERGENCE—A horizontal flow of water away from a common center, often associated with up-welling.

DOLDRUMS—The equatorial trough that is characterized by the light and variable nature of the winds.

DYNE—A force which, acting on a mass of one gram, imparts to that mass an acceleration of one centimeter per second per second.

The dyne is the unit of force of the *cgs system*.

EKMAN SPIRAL—A theoretical representation of the effect that a wind blowing steadily over an ocean of unlimited depth and extent and of uniform *viscosity* would cause the surface layer to drift at an angle of 45 degrees to the right of the wind direction in the Northern Hemisphere.

EQUATIONS OF MOTION—Specifically, Newton's laws of motion which relate force, velocity, and acceleration.

EQUATORIAL TIDES—Tides that occur approximately every two weeks when the moon is over the Equator. At these times, the moon produces minimum inequality between two successive *high waters* and two successive *low waters*.

EQUINOXES—The two points in the celestial sphere where the celestial equator intersects the ecliptic; also the times when the sun crosses the celestial equator at these points.

At this time day and night are of equal duration throughout the earth.

ESTUARY—(or *drowned river mouth, branching bay, firth, frith*). A tidal bay formed by submergence or drowning of the lower portion of a nonglaciated river valley and containing a measurable quantity of sea salt.

EURYHALINE—Adaptable to a wide range of salinity.

EURYTHERMAL—Tolerant of a wide range of temperature.

FETCH—(also called *generating area*). An area of the sea surface over which seas are generated by a wind having a constant direction and speed.

FJORD—(also spelled *fiord*). A narrow, deep, steep-walled *inlet* of the sea, formed either by the submergence of a mountainous coast or by entrance of the sea into a deeply excavated glacial trough after the melting away of the *glacier*.

FLOCCULATE—To aggregate into lumps, as when fine or colloidal clay particles in suspension in fresh water clump together upon contact with salt water and settle out of suspension; a common depositional process in estuaries.

FOOD WEB—The interaction of marine organisms with regard to the production, consumption, and decomposition of food in the sea. The interrelated sequences of organisms in which each is food for a higher member of the sequence.

FRINGING REEF—A *reef* attached directly to the shore of an island or continental landmass. Its outer margin is submerged and often consists of algal limestone, coral rock, and living coral.

FRUSTULE—The siliceous shell of a *diatom*, consisting of two valves, one overlapping the other. It is the principal constituent of *diatomaceous ooze*.

FULLY-DEVELOPED SEA—(also called *fully-arisen sea*). The maximum height to which ocean waves can be generated by a given wind force blowing over sufficient *fetch*, regardless of *duration*, as a result of all possible wave components in the spectrum being present with their maximum amount of spectral energy.

GEOLOGICAL OCEANOGRAPHY—The study of the floors and margins of the oceans, including description of submarine relief features, chemical and physical composition of bottom materials, interaction of sediments and rocks with air and sea water, and action of various forms of wave energy in the submarine crust of the earth.

GEOPHYSICS—The physics or nature of the earth. It deals with the composition and physical phenomena of the earth and its liquid and gaseous envelopes; it embraces the study of terrestrial magnetism, atmospheric electricity, and gravity; and it includes seismology, volcanology, oceanography, meteorology, and related sciences.

GEOSTROPHIC CURRENT—A current defined by assuming that an exact balance exists between the horizontal pressure gradient and the *coriolis force*.

GLAUCONITE—A green mineral, closely related to the micas and essentially a hydrous potassium iron silicate. Occurs in sediments of marine origin and is produced by the alteration of various other minerals in a marine reducing or *anaerobic* environment.

GRAM—A cgs unit of mass; originally defined as

the mass of 1 cubic centimeter of water at 4°C; but now taken as the one-thousandth part of the standard kilogram.

GRANITE—A crystalline plutonic rock consisting essentially of alkali feldspar and quartz. Granitic is a textural term applied to coarse and medium-grained granular *igneous rocks*.

GRAZING—The feeding of zooplanktonic organisms upon phytoplanktonic organisms. Generally in reference to the feeding of *copepods* upon *diatoms*.

GROIN—A low artificial wall-like structure of durable material extending from the land to seaward for a particular purpose.

HETEROTROPHIC NUTRITION—That process by which an organism utilizes only preformed organic compounds for its nutrition.

HOLOPLANKTON—Organisms living their complete life cycle in the floating state.

INTERNAL WAVE—A wave that occurs within a fluid whose density changes with depth.

INTERTIDAL—*see littoral*.

ION—An electrically charged group of atoms either negative or positive. The dissolved salts in seawater dissociate into ions.

ISLAND ARC—A term used for a group of islands usually having a curving archlike pattern, generally convex toward the open ocean, with a deep *trench* or *trough* on the convex side and usually enclosing a deep sea *basin* on the concave side.

ISOSTASY—The balance of large portions of the earth's crust as though they were floating in a denser medium.

ISOTOPE—A species of atom that differs from its ordinary counterpart by the number of neutrons in its nucleus and hence its mass number.

KNOT—A speed unit of one nautical mile (6,076.12 feet) per hour. It is equivalent to a speed of 1.688 feet per second or 51.4 centimeters per second.

LITTORAL—The benthic zone between high and low water marks.

LUNAR DAY—(or *tidal day*). The interval between two successive upper *transits* of the moon over a local meridian. The period of the mean lunar day, approximately 24.84 solar hours, is derived from the rotation of the earth on its axis relative to the movement of the moon about the earth.

LUNITIDAL INTERVAL—The interval between the moon's *transit* (upper or lower) over the local meridian and the following *high* or *low water*.

MAJOR CONSTITUENTS—Those chemical elements present in seawater which together make up over 99.9 percent of the known dissolved solid constituents of seawater. These include the following ions: chloride, sulfate, bicarbonate, bromide, fluoride, boric acid, sodium, magnesium, calcium, potassium, and strontium.

MANGROVE—One of several genera of tropical trees or shrubs which produce many prop roots and grow along protected low-lying coasts into shallow water.

MANTLE—1. The relatively plastic region between the *crust* and *core* of the earth. (Also called *asthenosphere*).
2. In biology, the body wall of some organisms, or the soft tissue next to the shell of mollusks.

MARINE ECOLOGY—The science which embraces all aspects of the interrelations of marine organisms and their environment and the interrelations between the organisms themselves.

MEGALOPLANKTON—Plankton larger than 1 cm.

MEROPLANKTON—Chiefly the floating development stages (eggs and larvae) of the benthos and nekton.

MICRON—(abbreviated μ). A unit of length equal to one-millionth of a *meter* or one-thousandth of a millimeter.

MIXED TIDE—A type of tide characterized by large inequalities in heights and/or durations of successive high and/or low waters.

MIXING—A general term that refers to the stirring or homogenation of seawater. Common mixing processes include wind, waves, and convection due to density instabilities or turbulence.

MOHOROVIČIĆ DISCONTINUITY—(abbreviated Moho). The sharp discontinuity in composition between the outer layer of the earth (the *crust*) and the next inner layer (the *mantle*). This was discovered by Mohorovičić from seismograms.

MONSOON—A name for seasonal winds derived from Arabic "mausim," a season. It was first applied to the winds over the Arabian Sea, which flow for six months from northeast and for six months from southwest, but it has been extended to similar winds in other parts of the world.

Mud—Pelagic or terrigenous detrital material consisting mostly of silt and clay-sized particles (less than 0.06 millimeter) but often containing varying amounts of sand and or organic materials. It is a general term applied to any sticky fine-grained sediment whose exact size classification has not been determined.

Nannoplankton—(or *centrifuge plankton*). Plankton within the size range 5 to 60 microns. Includes many *dinoflagellates* and smaller *diatoms*. Individuals will pass through most nets and usually are collected by centrifuging water samples.

Neap Tide—(or *neaps*). Tide of decreased range which occurs about every two weeks when the moon is in quadrature.

Nekton—Those animals of the *pelagic division* that are active swimmers, such as most of the adult squids, fishes, and marine mammals.

Nematocysts—The stinging mechanism of coelenterates, consisting of a chitinous sac filled with venom and pointed at one end. The pointed end can be everted by mechanical or chemical stimuli.

Neritic—The waters that overlie the continental shelf extending from low water level to the shelf break.

Nutrient—In the ocean any one of a number of inorganic or organic compounds or *ions* used primarily in the nutrition of primary producers. Nitrogen and phosphorus compounds are essential nutrients. Silicates are essential for the growth and development of *diatoms*. Vitamins such as B_{12} are essential to many *algae*.

Oceanic Province—The water filling the ocean basins seaward of the 100 m isobath. Often referred to as the "open ocean"

Oceanography—The study of the sea, embracing and integrating all knowledge pertaining to the sea's physical boundaries, the chemistry and physics of seawater, and marine biology.

Ooze—A fine-grained pelagic sediment containing undissolved sand- or silt-sized, calcareous or siliceous skeletal remains of small marine organisms in proportion of 30 percent or more, the remainder being clay-sized material.

Parasitism—A relationship between two species in which one lives on or in the body of its host, and obtains food from its tissues. Some authorities distinguish between a "commensal parasite," which obtains nourishment from its host without causing harm, and a "pathogenic parasite," which benefits at the expense of its host.

Pelagic Divison—A primary division of the sea which includes the whole mass of water. The division is made up of the neritic province which includes that water shallower than 100 fathoms, and the oceanic province, that water deeper than 100 fathoms.

Photic Zone—The layer of a body of water which receives ample sunlight for the photosynthetic processes of plants. The depth of this layer varies with the water's *extinction co-efficient*, the angle of incidence of the sunlight, length of day, and cloudiness; but it is usually 260 feet (80 meters) or more.

Photon—A quantity of electromagnetic energy whose value in *ergs* is the product of its frequency (v) in cycles per second and *Planck's constant* (h). The equation is: $E = hv$.

Photosynthesis—The manufacture of carbohydrate food from carbon dioxide and water in the presence of *chlorophyll*, by utilizing light energy and releasing oxygen.

Phytoplankton—The plant forms of *plankton*. They are the basic synthesizers of organic matter (by *photosynthesis*) in the *pelagic division*. The most abundant of the phytoplankton are the *diatoms*.

Plankton—The passively drifting or weakly swimming organisms in marine and fresh waters.

Primary Productivity—(or *gross primary production, primary production*). The amount of organic matter synthesized by organisms from inorganic substances in unit time in a unit volume of water or in a column of water of unit area cross section and extending from the surface to the bottom.

Progressive Wave—A wave which is manifested by the progressive movement of the wave form.

Pyncnocline—The vertical gradient of *density*.

Quadrature—The position in the phase cycle when the two principal tide producing bodies (moon and sun) are nearly at a right angle to the earth; the moon is then in quadrature in its first quarter or last quarter.

RADIANT ENERGY—(also called *radiation*). The energy of any type of *electromagnetic radiation*.

RADIOACTIVE DECAY—The disintegration of the nucleus of an unstable *isotope* by the spontaneous emission of charged particles and/or *photons*.

REFRACTION OF WATER WAVES—The process by which the direction of a wave moving in shallow water at an angle to the contours is changed. That part of the wave advancing in shallower water moves more slowly than the other part still advancing in deeper water, causing the *wave crest* to bend toward alignment with the underwater contours.

RESONANCE—The phenomenon of amplification of a *free wave* or oscillation by a forced wave or oscillation of exactly equal period.

RESPIRATION—An oxidation-reduction process by which chemically bound energy in food is transformed to other kinds of energy upon which certain processes in all living cells are dependent.

RIP CURRENT—The return flow of water piled up on shore by incoming waves and wind; a strong narrow surface current flowing away from the shore. A rip current consists of three parts: the *feeder* current flowing parallel to the shore inside the *breakers*; the *neck*, where the feeder currents converge and flow through the breakers in a narrow band or "*rip*"; and the *head*, where the current widens and slackens outside the breaker line.

SALINITY—A measure of the quantity of dissolved salts in seawater. It is formally defined as the total amount of dissolved solids in seawater in parts per thousand (0/00) by weight when all the carbonate has been converted to oxide, the bromide and iodide to chloride, and all organic matter is completely oxidized. These qualifications result from the chemical difficulty in drying the salts in seawater. In practice, salinity is not determined directly but is computed from *chlorinity, electrical conductivity*, refractive index, or some other property whose relationship to salinity is well established.

SAND—Loose material which consists of grains ranging between 0.0625 and 2.0000 millimeters in diameter.

SEAMOUNT—An approximately cone-shaped feature that rises at least 1000 m above the sea floor.

SEDIMENT—Particulate organic and inorganic matter which accumulates in a loose unconsolidated form. It may by chemically precipitated from solution, secreted by organisms, or transported by air, ice, wind, or water and deposited.

SEICHE—A *standing wave* oscillation of an enclosed or semienclosed water body that continues, pendulum fashion, after the cessation of the originating force, which may have been either seismic, atmospheric, or wave induced.

SEMIDIURNAL—*semidaily*. Actions that recur twice each twenty-four hours. When considering lunar tides, tides that recur twice each lunar day.

SESSILE—1. Attached directly by base, without stipe or stalk.

2. Permanently attached; not free to move about.

SHELF BREAK—(or *shelf edge*.) The line along which there is a marked increase of slope at the outer margin of a *continental shelf*.

SILL—A submarine ridge that separates two basins.

SILT—An unconsolidated sediment whose particles range in size from 0.0039 to 0.0625 millimeter in diameter (between clay and sand sizes).

SOLSTICE—One of the two points in the sun's orbit (the ecliptic) farthest from the celestial equator; the instant when the sun's declination is maximum.

SOLUBILITY—The extent to which a substance (solute) mixes with a liquid (solvent) to produce a homogeneous system (*solution*).

SPRING TIDE—Tide of increase range which occurs about every two weeks when the moon is new or full.

STANDING WAVE—A type of wave in which the surface of the water oscillates vertically between fixed points, called *nodes*, without progression. The points of maximum vertical rise and fall are called antinodes. At the nodes, the underlying water particles exhibit no vertical motion.

STENOHALINE—Capable of existence only within a narrow range of *salinity*.

STENOTHERMAL—Tolerant of only a very narrow range of temperature.

SUBLITTORAL—The benthic region that extends from the low water mark to the *shelf break*.

SUBMARINE CANYON—A relatively narrow, deep depression with steep slopes, the bottom of which grades continuously downward.

SUBTROPICAL HIGH—(or *subtropical anticyclone*,

oceanic anticyclone, oceanic high). One of the semipermanent highs of the subtropical high pressure belt. They lie over oceans, and are best developed in the summer season.

SUPRALITTORAL—The benthic zone above high tide level periodically moistened by waves and spray.

SURF ZONE—The area between the outermost *breaker* and the limit of wave *uprush*.

SWASH—(or *uprush, run-up*). The rush of water up onto the beach following the breaking of a wave.

SWELL—Ocean waves which have traveled out of their generating area. Swell characteristically exhibits a more regular and longer period and has flatter crests than waves within their *fetch*.

SYZYGY—The two points in the moon's orbit when the moon is in *conjunction* or *opposition* to the sun relative to the earth; time of new or full moon in the cycle of phases.

TABLEMOUNT—A *seamount* having a relatively smooth, flat summit.

TEST—The hard covering or supporting structure of many invertebrates, it may be enclosed within an outer layer of living tissue.

TIDAL CONSTITUENT—(also called *tidal component, partial tide*). One of the harmonic components comprising the tide at any point. The periods of the partial tides are derived from various combinations of the angular velocities of earth, sun, moon, and stars relative to each other.

TIDE—In the oceans, the periodic rise and fall of sea level due to the gravitational interaction of the sun, moon, and earth.

TIDE RANGE—The difference in height between consecutive *high* and *low waters*. Where the type of the tide is diurnal the mean range is the same as the diurnal range.

TRACTIVE FORCE—The horizontal resultant of the tide-producing force over the earth.

TRADE WINDS—The wind system, occupying most of the tropics, which blows from the *subtropical highs* toward the equatorial trough: a major component of the general circulation of the atmosphere.

TRENCH—A long, narrow and deep depression of the sea floor, with relatively steep sides.

TROPHIC LEVEL—A successive stage of nourishment as represented by links of the food chain. Primary producers (*phytoplankton*) constitute

the first trophic level, herbivorous *zooplankton* the second trophic level, and carnivorous organisms the third trophic level.

TROPIC TIDE—The tide that occurs twice monthly when the effect of the moon's maximum declination north or south of the Equator is greatest.

T-S CURVE—The plot of temperature versus salinity data of a water column. The result is a diagram which identifies the water masses within the column, the density, and the stability of the column.

TSUNAMI—(or *tunami, tidal wave, seismic sea wave*). A long-period sea wave produced by a submarine earthquake or volcanic eruption. It may travel unnoticed across the ocean for thousands of miles from its point of origin and builds up to great heights over shoal water.

TURBIDITY CURRENT—(or *density current*). A highly turbid, relatively dense current carrying large quantities of clay, silt, and sand in suspension which flows down a submarine slope through less dense seawater.

UPWELLING—The process by which water rises from a lower to a higher depth, usually as a result of *divergence* and offshore currents.

Upwelling is most prominent where persistent wind blows parallel to a coastline so that the resultant wind-driven current sets away from the coast. It constitutes a distinct climatogenetic influence by bringing colder water to the surface.

The upwelled water, besides being cooler, is richer in plant nutrients, so that regions of upwelling are generally also regions of rich fisheries.

VARVE—A sedimentary deposit, bed, or lamination deposited in one season. It is usually distinguished by color or composition and used as an index to changes in the depositional environment.

VISCOSITY—(or *internal friction*). That molecular property of a fluid which enables it to support *tangential stresses* for a finite time and thus to resist deformation.

WAVE—1. A disturbance which moves through or over the surface of the medium (here, the ocean), with speed dependent upon the properties of the medium.

2. A ridge, deformation, or undulation of the surface of a liquid.

WEST WIND DRIFT—(sometimes called *Antarctic Circumpolar Current*). The ocean current with the largest volume transport (approximately 110 x 10⁶ cubic centimeters per second); it flows from west to east around the Antarctic continent and is formed partly by the strong westerly wind in this region and partly by density differences.

WIND-DRIVEN CURRENT—(sometimes called *wind drift, drift current*). A current formed by the force of the wind. Theoretically, currents produced by the wind will set to the right of the direction of the wind in the Northern Hemisphere and to the left in the Southern Hemisphere.

WIND SET-UP—The vertical rise in the *still water level* on the leeward side of a body of water caused by wind stresses on the surface of the water.

WIND-WAVE—Wave motion that results from the interaction of the wind and the sea surface.

ZOOPLANKTON—The animal forms of *plankton*. They include various crustaceans, such as copepods and euphausiids, jellyfishes, certain protozoans, worms, mollusks, and the eggs and larvae of benthic and nektonic animals. They are the principal consumers of the *phytoplankton* and, in turn, are the principal food for a large number of squids, fishes, and baleen whales.

index

index

Absolute temperature scale, 305
Abyssal environment, 225, 317
Abyssal hills, 25
Abyssal plains, 27
Abyssal sediments (*see also* Marine sediments), 263
Acceleration:
  definition, 299, 300, 317
  of gravity, 125, 126, 300
Accessory minerals in igneous rocks, 313
Acidic solution, 312
Adsorption of ions on clay minerals, 61
Advection, 66, 317
Aegean Sea, 253
Aerobic (*see also* Bacteria), 317
Akutan Pass, 157
Aleutian Islands, 12, 52
Algae:
  blue-green, 192
  brown, 194
  and coastal erosion, 181
  definition, 192, 317
  encrusting, 193, 194
  green, 192
  red, 193, 194
Allogenic sediment, 255, 317
Alluvial fan, 23

Amazon River, 157
Amphibians, 214
Amphidromic Point, 155, 317
Amphineura:
  figure, 204
  synopsis, 202
Anaerobic (*see also* Bacteria), 317
Anchovy:
  in the food chain, 241
  protection from enemies, 219
Andaman Sea, 12
Andésite line, 30, 32
Angler fish, 218, 234, 235
Angular rotation of the earth, 89
Anion:
  definition, 312
  exchange processes, 62
Annelida:
  figure, 203
  synopsis, 205
Anoxic water, 188
Antarctic Bottom Water, 110, 116, 117
Antarctic Circle, 137
Antarctic Convergence, 98, 109, 317
Antarctic Intermediate Water, 113–115
Anthozoa (*see also* Coelenterata):

Anthozoa (*cont.*)
  figure, 200
  synopsis, 201
Aphelion, 146
Aphotic zone, 219, 237
Apparent oxygen utilization, 67
Arabian Sea, 12
Aragonite in marine sediments, 253
Archimedes principle, 298
Arctic Circle, 137
Arctic Convergence, 98, 109, 111
Argentine Basin, sedimentation in, 268
Arrow worm (*see* Chaetognatha)
Arthropoda:
  figures, 206
  synopsis, 205
Ascension Island, 28
Atlantis, legend of, 253
Atmospheric detritus, 86
Atoll, 317
  classification of reefs, 24, 25
  and coralline algae, 192
  origin of, 32, 33
Atoms, basic structure, 309
Authigenic sediments, 255, 257, 317
Autoanalyzer, 284, 287
Autotrophic nutrition, 216, 318

Autumnal equinox, 137
Auxins in seawater, 59
Auxospore (see Diatom)
Aves, 210
Azoic zone, 225
Azore Islands, 28, 46

Bacillariophyceae(see also Diatom):
  figure, 195
  synopsis, 194
Backwash, 163
Bacteria:
  aerobic, 237
  anaerobic, 188, 238
  decomposition of organic matter, 237, 238
  distribution in the sea, 237
  regulation of seawater composition, 62, 212, 217
  remineralization of nutrient elements, 237, 241
  sulfate reducing, 68
  and trophic levels, 239
Baleen (see also Whale), 205, 209, 210, 318
Baltic Sea, 11
Bar-built estuaries, 183, 188
Barite in marine sediments, 255
Barnacle:
  acorn, 206
  benthic adaptations, 223, 231, 232
  goose, 206
  larvae, 205, 206
  nauplius, 229
Barrier beach, 315
Barrier island, 188, 189
Barrier reef, 24, 25, 33, 318
Barycenter, 137, 139
Basalt, 16, 180, 318
Basaltic glass, 253, 254
Basic solution, 312
Bathypelagic organisms, 234, 235, 318
Bathyscaphe, 274
Bay of Bengal, 12
Bay of Biscay, 12
Bay of Fundy, 135, 153, 158
Beach, 318
  berm, 169–171
  drift of sediment, 167, 168
  nearshore circulation on, 160–176
  profile, 169, 170
Benthic environment, 318
  adaptations for survival in, 231, 232

Benthic environment (cont.)
  classification of, 221, 222
  organisms in the, 223
Benthos:
  adaptations of, 230–232
  deep-sea, 224, 225
  environment of, 221
  organisms, 223
  and sediment layering, 263
Bering Sea, 12
Berm, 169–171, 318
Biological components of sediments, 197, 248, 251, 252, 257
Biological oceanography, 318
Bioluminescence, 318
  in Ctenophora, 202
  in fish, 235
Biotic unit, 219
Black Sea, 68, 77, 188, 239
Blake Plateau, Florida, 20
Blue-green algae, 192
Bottom-water, 110, 111
  mixing cycle of, 116
  North Atlantic, 115
Boundary between ocean basins, 11
Box samples, 275, 276
Brachiopod, 212
Brackish seawater, 74
Breaking wave, 163
Brittle star (see Echinodermata)
Brown algae:
  adaptation, 232
  Fucus, 191
  kelp, 194
  Neriocystis, 193, 194
  Sargassum, 191, 194
Brown clay, 318
  classification of, 257
  deposition rate, 268
  layering, 264
  organic content, 225
  origin, 259, 260
Buccoo Reef, West Indies, 181
Buffer, 318
  carbonate system, 70
  silica-clay system, 63
Bulk properties of sediments, 246, 247
Bungo Strait, 157
Buoyancy, 298
By-the-wind-sailor (see Velella)

Calanus (see also Copepoda), 206
  vertical migrations, 228–231
Calcareous ooze, 257, 259, 260

Calcium carbonate:
  buffer system, 70
  deposition associated with glacial stages, 265
  ooze, 263, 264
  precipitation of, 261
  reef (see also Coral reefs), 24, 32
Calorie, 305, 306
Capacity of sediment transport, 167
Cape Horn, 11
Cape of Good Hope, 11
Cape Verde Islands, 46
Capillary waves (see also Waves), 119, 121, 122, 318
Carbohydrate, synthesis of, 235, 236
Carbonate (see Calcium carbonate)
Carbon dioxide:
  photosynthesis, 236
  respiration, 216
  in seawater, 70
Carbon-14 dating technique, 267, 318
Caribbean Sea, 11, 194
Carlsberg Ridge, 28
Carnivore, 217
Caspian Sea, 12
Cation:
  definition, 312
  exchange process, 61
Cell division (see also Diatom), 194, 196
Celsius temperature scale, 304
Centigrade temperature scale, 304
Central Water Mass, 98, 110–114
Centrifugal force, 137, 139
Cephalopoda:
  figure, 204
  synopsis, 202
C/G/S system of measurements, 293, 318
Chaetognatha:
  arrow worm, 207
  bathypelagic, 234
  figure, 207
  Sagitta, 207
  synopsis, 207
Challenger, H.M.S., 3, 4, 65, 318
Challenger Deep, Mariana Trench, 29
Chambered Nautilus (see also Cephalopoda), 202, 204
Channel Islands of California, 52
Chart datum, 152, 318
Chemical components in marine sediments, 248, 253–257
Chemical elements, 309

Chemical energy, 306
Chemical nomenclature, 309–313
Chemical oceanography, 318
Chemical symbols, 311
Chert, 255, 263, 264, 318
Chesapeake Bay, 135
Chiton (*see* Amphineura), 204
Chlorinity, 65, 66, 71, 72, 318
Chlorophyll, 235, 319
Chlorophyta (*see also* Green algae), 192
Chondrichthyes, 210
Chordata:
    figure, 209
    synopsis, 207
Cilia, 202
Circumpolar Water, 98
Cirripedia (*see also* Barnacle):
    figure, 206
    synopsis, 205
Clam (*see* Pelecypoda)
Clark-Bumpus sampler, 288, 290
Clay, 319
    minerals in marine sediments, 253, 255, 257–260, 264
    sized particles, 249
    as a weathering product of igneous rocks, 248
Climatic control of deep-sea sedimentation, 265, 266
Cnidaria (*see* Coelenterata)
Coastal-plain estuary (*see also* Estuary), 183
Coastal vegetation:
    as a beach stabilizer, 170, 181, 182
    mangrove, 182, 197
    Zostera, 182
Cobb Seamount, 30, 31
Coccolith (*see also* Coccolithophore), 319
Coccolith ooze, 257, 259, 260, 264
Coccolithophore:
    coccolith, 196
    and color of the sea, 85
    figure, 195
    in marine sediments, 257, 259, 260, 264
    rabdolith, 196
Cod fish:
    food chain, 241
    as nekton, 222, 223
    protection from enemies, 219
Coelenterata:
    coral, 200, 201
    figures, 200
    hydroid polyp, 200, 201
    jellyfish, 201, 222, 228, 230
    medusa, 200, 201
    nematocyst, 201

Coelenterata (*cont.*)
    Physalia, 209, 228
    sea anemone, 200, 201
    sea wasp, 201
    synopsis, 201
    Velella, 228, 230
Colloids, 59, 311
Color:
    of marine sediments, 261, 262
    of the sea, 84, 85
Colorimetry, 283, 284
Comb jelly (*see also* Ctenophora), 201
Commensalism, 218, 319
Compaction of sediments, 247
Competency of sediment transport, 165, 167
Condensation hypothesis of the earth's origin, 15
Conductivity of seawater, 283
Conservation of energy, 297
Continental crust, 16, 49
Continental drift, 35–37, 49, 51
Continental margin, 19, 20, 24, 51–53
Continental rise, 19, 21, 23, 260, 319
Continental shelf, 19–21, 51–53, 221, 268, 269, 319
Continental slope, 19–21, 319
Continuous plankton sample, 291
Convection, 319
    in the earth, 43
Convergence, 103, 104, 107, 111, 319
    Antarctic, 98, 99, 111
    Arctic, 98, 109
    of longshore currents, 163
Cook Inlet, Alaska, 159
Copepoda (*see also* Calanus):
    adaptations, 229
    figures, 206
    as grazers, 205
    protection from enemies, 219
    synopsis, 205, 206
    vertical migration, 230, 231
Coral (*see also* Coelenterata):
    benthic adaptation, 223
    figures, 200
Corallina (*see also* Red algae):
    figure, 251
    as a sediment producer, 251
Coral reef, 319
    classification, 24
    organisms producing, 193, 194
    origin, 32, 33
Core of the earth, 14, 15
Coring tube, 274, 275
Coriolis force, 89–100, 105, 106, 108–110, 319

Cotidal line, 155, 157
Countercurrent (*see* Equatorial Countercurrent)
Crabs (*see also* Decapoda), 205
    regulatory mechanisms for body fluid salinity, 215
Crustacea:
    bathypelagic, 234
    figures, 206
    as nekton, 222
    as plankton, 222
    synopsis, 205
Crustal plates, 49–51
Crust of the earth, 319
    continental, 14–16
    magnetic polarity, 37–39, 49
    movement of, 49–51
    oceanic, 14–16, 46, 49
Ctenophora:
    figure, 201
    synopsis, 202
Currents, ocean:
    cause of deep water currents, 112
    cause of surface currents, 98–108
    meridonal circulation, 108
    surface, 99
Cyanophyta (*see* Blue-green algae)

Daily migration of zooplankton, 228–230
Darwin, Charles, 24, 33
Dating marine sediments, 266–268
Daytona Beach, Florida, 246, 261, 262
Decapoda, 205
    adaptation, 229
Declination, 319
    of the moon, 141–143
Decomposition of organic matter, 237
DDT, 7
Deep-sea cores, 49
Deep-sea drilling, 27, 35, 47, 51, 278
Deep-sea environment, 221
Deep-sea floor:
    features of, 25
Deep-sea sediments (*see also* Marine sediments), 258
Deep-sea trenches, 28–30, 43, 49
Deep-water masses, 110, 115, 116
Deep-water waves (*see also* Waves), 120, 125, 319
Density, 319
    definition of, 297
    of earth features, 15, 16

Density (cont.)
  of marine organisms, 227, 228
  profile, 77, 80, 107
  relationships of seawater, 74–78
Dentrification, 69, 238, 319
Depth of frictional resistance, 101
Deseado River, Argentina, 158
Desiccation, 319
  adaptations to avoid, 231, 232
  of marine organisms, 214
Detritus, 319
Diagenetic components of marine
    sediments, 255, 319
Diapir Barriers, 52
Diatom:
  annual cycle of, 242
  auxospore, 196
  cell division, 194, 196
  figures, 195
  frustule, 194–196, 251
  in marine sediments, 251, 257,
    258, 263
  resting spore, 196, 231
Diatom ooze, 257, 259, 260, 263
Diatoms in marine sediments, 251,
    257
Diffusion, 66, 319
Dinoflagellate, 194:
  color of the sea, 85
  flagella, 196
  Gonyaulax, 195
  red tides, 85, 195
Dinophyceae (see also Dinoflagel-
    late):
  synopsis, 196
Dipole, 56, 319
Discoasters in marine sediment,
    251
Dispersion in waves, 126, 127
Dittmar, C. R., 65
Diurnal tide, 134, 319
Divergence, 103, 104, 107, 319
Doldrums, 91, 106, 107, 320
Dolomite in marine rocks, 253
Dredge, 274, 275, 287
Drift bottle, 285
Drowned-river valley, 183
Dynamic equilibrium, 301
Dyne, 300, 320

Earth:
  core, 14
  crust, 15
  dimensions, 10
  distance from sun, 136
  mantle, 14
  orbit around sun, 136
  origin, 15

Earth (cont.)
  physical characteristics, 15
  shape, 9
Earthquake epicenters, 41–43
East China Sea, 12
East Pacific Rise, 28, 47, 49
Ebb tide, 133, 154
Echinodermata:
  brittle star, 207
  figures, 208
  floating adaptation, 228
  sand dollar, 207
  sea cucumber, 207
  sea urchin, 207
  starfish, 207
  synopsis, 207
Echo sounder, 271, 273
Ecliptic, 141, 142, 144
Ecology (see also Marine ecology),
    226
Eel grass (see also Zostera), 182,
    197
Ekman, V. W., 100, 105, 130
Ekman current meter, 285, 288
Ekman spiral, 100, 101, 320
Ekman transport, 101–105
Electrical energy, 296
Electrolyte, 312
Electromagnetic spectrum, 309
Electron, 209
El Niño, 223
Embryophyta:
  synopsis, 192
Encrusting algae, 193
Endoskeleton, 219
Energy, forms of, 296
Equation of motion, 299, 320
Equatorial Countercurrent, 98,
    106, 223
Equatorial currents, 106–108
Equatorial tide, 136, 142, 143,
    320
Equilibrium theory of tides, 145
Equinox, 137, 320
Eratosthenes, 11
Erosion:
  coastal, 178, 181
  of continental rocks, 248, 249
Erratic sediments, 250
Estuaries, 320
  biological productivity, 189
  circulation, 185–189
  sedimentation, 185, 261
  sediment types in, 261
  types of, 183
Ethiopia, 32
Eukaryotes, 190
Euphausiacea:
  figure, 206
  synopsis, 205

Euphausia pacifica (see also Eu-
    phausiacea), 206
Euryhaline, 231, 320
Eurythermal, 231, 320
Evaporation:
  related to atmospheric circula-
    tion, 94
  and surface circulation, 107
Exoskeleton, 219
Extraterrestrial components of sedi-
    ments, 248, 252, 260

Fahrenheit temperature scale, 304
Faroe Islands, 46
Fathom, 294
Feldspar, 313
Ferrel's Law, 90
Ferromagnesian mineral, 313
Fetch, 121, 122, 320
Fisheries:
  Peru anchovy, 241
  world, 6
Fishes (see also under specific
    fishes):
  adaptations, 232
  angler fish, 219
  bathypelagic, 234
  cod fish, 222, 223, 241
  endoskeleton, 219
  food chain, 241
  as parasites, 218
  protection from enemies, 219
  respiration of, 216
  synopsis, 210
Fish protein concentrate, 6
Fjords, 320
  anoxic conditions, 188, 239
  characteristics, 185, 186
  circulation, 187
  as a type of estuary, 183
Flagella (see Dinoflagellate)
Flatworms (see Platyhelminthes)
Floating adaptations:
  phytoplankton, 227, 228
  zooplankton, 228, 229
Flocculation, 320
  in estuaries, 185, 261
Flood tide, 133, 154
Food chain (see Food web)
Food resources in the ocean, 6
Food web, 235–239, 320
  theoretical models, 244, 245
Foraminifera, 199
  in marine sediments, 217, 251,
    268
Force:
  centrifugal, 137–139
  Coriolis, 89, 99, 100, 105, 106,
    108, 110

Force (*cont.*)
  definition, 299
  diagrams, 301
  friction, 99, 108
  gravity, 100, 105, 110, 137, 138, 300
  primary, 99
  resolution procedure, 301
  secondary, 99
  vectors, 302
Forced waves, 144
Forchammer, J. G., 64
Fracture zones, 28, 49
Freezing of seawater, 74–76
Freshwater lid effect, 77
Friction:
  force, 99, 108
  viscosity and, 79, 131
Fringing reef, 24, 25, 33, 181, 320
Frustule (*see also* Diatom), 320
Fucus (*see* Brown algae)
Fully-developed sea, 320

Gases dissolved in seawater, 66
Gastropoda:
  figures, 204
  littorinid snail, 232
  in marine sediments, 250, 257
  synopsis, 202
Geological oceanography, 320
Geologic time scale, 213, 295
Geomagnetism, 37
Geophysics, 320
Geostrophic current, 110, 320
Gills in marine organisms, 216
Glacial marine sediments, 260, 261
Glaciation (*see also* Pleistocene Epoch), 11
Glauconite in marine sediments, 255, 320
Global tectonics, 34
Globigerina, 198, 199
  deposition rate, 268
  in marine sediments, 251, 256–258
Globigerina ooze, 257, 259, 260
  deposition rate, 268
Glomar Challenger, 47, 264, 278
Glucose, 236
Gonyaulax (*see* Dinoflagellate)
Grab sampler, 274, 275, 287
Grading of marine sediments, 249, 250, 263
Gram, 294, 320
Grand Banks earthquake, 23
Granite, 16, 313, 321
Gravitational constant, 138, 300

Gravitational force, 105, 106, 110, 137, 138, 300
Gravity core, 275–277
Gravity measurements at sea, 279
Grazing, 321
  by copepods, 205
  by zooplankton, 239, 242
Great Barrier Reef of Australia, 24
Green algae:
  Halimeda, 192, 251
  sea lettuce, 192
  Ulva, 191, 192
Groin, 175, 176, 321
Gulf of California, 12, 189
Gulf of Mexico, 12
  salt domes, 52
Gulf of St. Lawrence, 12
Gulf Stream, 20, 96, 108, 115, 223
Gulf weed (*see* Brown algae)
Guyot (*see also* Tablemount), 30, 32

Hadley cells, 92
Half-meter net, 287, 289
Halimeda (*see* Green algae)
Harmonic analysis, 147, 149
Harvest of fish (annual), 6, 239
Hawaii:
  beach sands, 262
  islands, 47
Heat capacity, 82, 215, 304–306
Heat exchange across the sea surface, 82
Heat flow, 42–44
Herbivore, 217
Hermit crabs (*see* Decapoda)
Herring:
  food web, 239, 240
Heterotrophic nutrition, 216, 217, 321
History of the ocean basins from deep-sea drilling, 263–265
Holdfast, 232
Holocene Epoch, 265, 266
Holocene marine transgression, 178, 180, 183
Holoplankton, 222, 321
Holothurian (*see also* Echinodermata):
  floating adaptation, 229
Holozoic nutrition, 217
Horse latitudes, 91
Hudson Bay, 11, 12
Humic substances in seawater, 59
Hydration, 56
Hydrogen sulfide:
  anoxic conditions, 68, 188
  in Black Sea, 68, 187

Hydrogen sulfide (*cont.*)
  at bottom of basins, 68, 187
  exclusion of animals by, 68, 188
  in sediments, 262
  and sulfate reduction, 238, 239
Hydroid polyp (*see also* Coelenterata), 200
Hydrostatic pressure, 74
Hydrozoa (*see also* Coelenterata):
  figure, 200
  synopsis, 201
Hypsographic curve, 13, 14

Ice age (*see also* Pleistocene Epoch), 8
Iceland, 28
Igneous rocks:
  composition of, 310
  definition, 313
Indian spring low water, 152
Intermediate Water Mass:
  characteristics, 112–115
  origin, 112–115
  position in the water column, 110, 115
Internal waves, 129, 130, 321
Interstitial water of sediments, 247, 255
Intertidal zone, 321
  classification of, 221
  organisms in, 231, 232
Ion, 321
  definition, 311, 313
  exchange processes in desalination, 5
  pair, 57, 59
Island, 30, 32, 46–48
Island arc, 12, 43, 321
Isostasy, 17, 49, 176, 321
Isotope, 321
  definition, 309
  of oxygen, 68, 69, 266

Japan, Islands of, 12, 52
Japan Trench, 28, 30
Java Trench, 12
Jellyfish (*see also* Coelenterata), 200

Katmai eruption, 252
Kelp (*see* Brown algae)
Kelvin temperature scale, 305
Kerogen in marine sediment, 59
Kinetic energy, 296, 300
Knolls, 30
Knot, 155, 321
Krakatoa:
  ash, 252
  explosion, 128

Krill (*see* Euphausiacea)
Kurile Trench, 30
Kuroshio Current, 96

Lake Superior, 135
Lamellibranchia (*see* Pelecypoda)
Landslide surge, 132
Latent heat energy of evaporation, 82
   definition, 307, 308
Latent heat energy of fusion, 82
   definition, 306
Laws of motion, 88, 89, 138, 299
Libya, 82
Light penetration in the sea, 83
Limpet (*see also* Gastropoda), 204
Linnaeus, Karl von, 191
Lipid, production of, 59
Lisbon, Portugal, 150
Lithothamnion (*see* Red algae)
Littoral, 321
   currents, 163–165
   drift of sediment, 165, 168, 172, 176
   environment, 218–221, 223
   organisms, 231
Littorinid snail (*see also* Gastropoda), 232
Lituya Bay, Alaska, 132
Lobsters (*see also* Decapoda):
   phyllosoma larvae, 229
Longhurst-Hardy plankton sampler, 290, 291
Longshore current, 162, 163
Longshore drift of sediment, 165
Lunar day, 141, 318
Lunar fortnightly tidal constituent, 145
Lunar month, 138, 141
Lunar semidiurnal tide, 141, 147
Lunitidal interval, 144, 321

Magellan, 270
Magnetic polarity of the crust, 37–39, 49, 267, 268
Magnetite, 37
Major constituents of seawater (*see also* Seawater), 57, 321
Mammalia, 210
   as nekton, 222
Manganese nodules, 6, 253–257
Mangrove, 182, 197, 321
Manila, R. P., 134
Mantle, 321
   of the earth, 14, 15, 49
   of mollusks, 202
Mariana trench, 12, 29

Marine ecology, 321
Marine sediments:
   accumulation rates, 268
   calcareous, 260, 261
   classification, 248, 257
   climatic control of, 265, 266
   color, 261, 262
   dating of, 266, 267
   distribution of, 259, 260
   grading, 249, 250, 263
   layering, 263, 264
   organic content, 251
   and productivity, 265
   rafting, 250
   settling velocity, 249
   size classification, 249
   thickness of, 268
   transportation, 248, 249
   types of, 257
   weathering, 248
Marshall Islands, 32
Mass distribution current, 110
Mean lower low water, 152
Mean low water, 152
Mechanical energy, 296
Mediterranean Sea, 11, 12, 189
Mediterranean Water, 115
Medusa (*see* Coelenterata)
Megaloplankton, 222, 321
Mendocino Fracture Zone, 28
Meroplankton, 222, 321
Metamorphic rocks, 313
Meteor expedition, 273
Meter, unit of length, 294
Metric system of measurements, 293
Mica, 313
Micron, 293, 321
Mid-Atlantic Ridge, 27, 28, 45, 46, 49, 51
Mineral:
   definition, 313
   in igneous rocks, 310
Mississippi River, 184
Mixed layer, 241–245
Mixed tides, 147, 153, 321
Mixing, 321
   time for the oceans, 2, 65, 116
   by winds, 241–245
Mixture (chemical definition), 310
Mohorovičić discontinuity, 16, 321
Molecular diffusion, 241
Molecule, 309, 310
Mollusca, 202
   figures, 204
   synopsis, 202
Monazite, 262
Monsoon, 92, 93, 98, 106, 321
Montmorillonite, 253

Mornington Peninsula, Australia, 180
Mud, 322
   as a sediment type, 257
Murray Fracture Zones, 28
Mussels (*see* Pelecypoda)

Nannoplankton, 222, 291, 322
Nansen bottle, 281, 283
Natural period of oscillation, 153
Nautical mile, 294
Neap tide, 134, 144, 145, 322
Nearshore circulation, 160
Nearshore sediment transport, 165
Nekton, 322
   adaptations, 232–235
   deep-sea, 233
   organisms, 222, 223
Nematocysts (*see also* Coelenterata), 322
Nereocystis (*see* Brown algae)
Neritic province, 322
   classification of, 220, 221
   productivity in, 245
   sediment type in, 256
Neutron, 309
Newton, Sir Isaac (*see also* Laws of motion), 88, 89, 138, 145
Nitrate:
   annual cycle of, 241, 242
   measurement of, 284
   as a nutrient ion, 59
   reduction and dentrification, 69, 70, 237–239
   regeneration of, 237
Nitrogen dissolved in seawater, 69
Noble gases in seawater, 71
North Atlantic Deep Water, 115
North Sea, 11
Norwegian Sea, 225
Notochord, 207
Nudibranch (*see also* Gastropoda), 204
Nutrients, 322
   analysis for, 284, 285
   annual cycle of, 241, 242
   and photosynthesis, 235, 236, 241
   regeneration of, 237
   in seawater, 59
Nutrition, 216–218

Ocean:
   area, 10
   as a natural resource, 4–6
Ocean Basins:
   area, 12
   characteristics, 11, 12
   depth, 13

Ocean currents:
  mass transport, 106
  mixing rates, 2, 65, 116
  speed, 106, 108, 116
Oceanic crust, 16, 46, 49
Oceanic province, 322
  classification of, 220, 221
  productivity in, 243–245
Oceanic ridges, 27, 43, 49
Oceanography, 1, 322
Octopus (see also Cephalopoda), 204
Offshore bars, 170
Oikopleura (see Tunicata)
Omnivore, 217
Ooze (see also Marine sediments), 322
  deep-sea, 225
Orbital motion in waves, 124–126
Organic compounds, 59
Orthogonal, 162
Osmosis, 5, 214
Osteichthyes (see Fish)
Overturn:
  and density instability, 78
Oxygen:
  exchange processes in organisms, 216
  isotopes of, 68, 69
  minimum zone, 68
  and photosynthesis, 235, 236
  in seawater, 66, 67
Oysters (see Pelecypoda)

Pacific Equatorial Undercurrent, 108
Palagonite, 253, 255
Paleoclimate, 265, 266
Palmer Peninsula, 11
Pangaea, 35, 51
Parasitism, 217, 218, 322
Partial tide, 147, 150
Pelagic environment, 220, 320
Pelagic sediment (see also Marine sediments):
  classification, 255, 257
  color, 262
  composition, 260
  depth range, 259, 260
  distribution, 258
  layering, 263
  rate of deposition, 268
Pelecypoda:
  figures, 204
  in marine sediments, 251
  synopsis, 202
Penzhinskaya Bay, Siberia, 159
Perihelion, 146
Pernambuco, Brazil, 150
Persian Gulf, 12, 82

Peru-Chile Trench, 28, 51
Phaeophyta (see also Brown algae):
  synopsis, 194
Phillipsite, 253–255
Phosphate (see also Nutrients):
  measurement of, 284
Phosphorite, 255
Photic zone, 85, 219, 220, 227, 237, 238, 241, 242, 322
Photon energy, 308, 322
Photophores, 235
Photosynthesis, 322
  and the carbon dioxide cycle, 70
  as a chemical reaction, 235
  and the development of life in the sea, 212
  measurement of, 291, 292
  and the oxygen cycle, 63, 67
  and radiant energy, 84
  requirements for, 216
Phyllospadix (surf grass), 197
Physalia (see also Coelenterata), 201, 228
Phytoplankton, 322
  annual growth cycle, 241–245
  chemical composition, 236
  coccolithophores, 196
  definition, 222
  density, 227
  diatoms, 194
  dinoflagellates, 196
  in estuaries, 189
  floating adaptations, 227
  as food for zooplankton, 205
  nutrition, 227
  sinking rate, 227
  size, 227
  and upwelling, 104
Piston core, 276–278
Plankton, 322
  protozoan, 197–199
  synopsis, 222, 223
Plankton pump, 288
Plate tectonics, 49–52
Platyhelminthes:
  figure, 203
  synopsis, 202
Pleistocene Epoch, 11, 17, 21, 23, 52, 176, 183, 265, 266, 269
Polar easterlies, 91
Polar Front, 91, 101, 109
Polychaeta:
  figure, 203
  synopsis, 205
Polysiphonia (see Red algae)
Porifera (see also Sponge):
  figure, 199
  synopsis, 199

Porolithon (see also Red algae):
  in marine sediments, 251
Porpoise, 209, 210
Portuguese-Man-of-War (see Physalia)
Potassium-40 technique for dating igneous rocks, 267
Potential energy, 296, 301, 303
Prawns (see Decapoda)
Precipitation:
  and atmospheric circulation, 94
  and monsoons, 92, 93
  and surface circulation, 107
Pressure, 73, 74, 76, 302
Pressure transducer to measure waves and tides, 286, 289
Primary productivity, 322
  annual growth of phytoplankton, 240
  measurement of, 291, 292
  in the neritic province, 244, 245
  in the oceanic province, 243–245
  and photosynthesis, 235
  polar regions, 243
  and sedimentation, 265
  temperate regions, 242
  theoretical models of, 244, 245
  tropical regions, 242
Progressive waves, 119, 120, 124, 153, 322
Progressive wave tide, 153, 154
Prokaryotes, 190
Protista, 191
Proton, 309
Protoplasm, production of, 236
  in marine sediments, 251
Protozoa:
  figure, 198
  synopsis, 197, 199
  test, 197–199, 251
Pteropod Ooze, 257, 259, 260
Pteropods (see also Gastropoda):
  in sediments, 251, 257
Puget Sound, 77, 79, 135, 150
Pycnocline, 106, 129, 322
  seasonal change in, 241–245
Pyrite, 253, 256

Quadrature, 144, 322
Quantum (energy), 308
Quartz, 313

Rabdolith (see Coccolithophore)
Radiant energy, 323
  and the electromagnetic spectrum, 308
  as a form of energy, 296

Radiant energy (*cont.*)
  in the ocean, 82–85
  and photosynthesis, 235
  from the sun, 82
Radioactive decay, 323
  and atomic structure, 309
  and rock dating, 266, 267
Radiolarians:
  figure, 198
  in marine sediment, 251, 257
  ooze, 257, 259, 260
  synopsis, 198, 199
Radula, 202
Rafting of sediments, 250
Rance Estuary, France, 158, 159
Ray (*see also* Chondrichthyes),
  210
Red algae:
  Corallina, 251
  encrusting, 193
  Lithothamnion, 192
  Polysiphonia, 191
  Porolithon, 251
Red Sea, 12, 189
  warm brines, 256
  water type, 115
Red tides (*see also* Dinoflagellate),
  85, 195
Reef:
  atoll, 24, 32, 33, 192
  barrier, 24, 33
  fringing, 24, 33, 181
Reef barriers, 52
Refraction of waves, 161–163, 175,
  320
Relict sediments, 52, 268
Remineralization of organic mat-
  ter, 237
Reptilia:
  figure, 209
  synopsis, 210
Residence time, 62
  of surface water, 116
Resonance and wave amplitude,
  131, 323
Respiration, 216, 235, 236, 323
Resting spores (*see* Diatom)
Rhodophyta (*see also* Red algae):
  synopsis, 194
Rift valley, 28
Rip current, 163–165, 169, 323
Rock (defined), 313
Rocky Mountains as compared to
  deep sea topography, 27
Rule of constant proportions, 64,
  65

Sagitta (*see also* Chaetognatha),
  207

St. George, Bermuda, 150
Salinity, 321
  bridge, 72, 283
  changes of, 74
  definition, 64–66
  ecological factor, 214
  measurement of, 72, 283
  origin, 60–64
  ranges in the littoral environ-
    ment, 221
  and stability, 77
  surface values, 94
  variations of, 64, 94
Salinometer, 72, 284
Salt (definition), 312
Salt dome, 52
Salt-wedge estuary, 184, 189
Sand, 323
  beach, 172
  dunes, 170, 172
  particle size, 249
  sorting by density, 171
Sand crabs (*see also* Decapoda):
  planktonic larval period, 222
Sand dollar (*see* Echinodermata)
San Francisco Bay, 157, 189
San Jose River, Argentina, 158
Santorini Volcano, 253
Saprozoic nutrition, 217
Sargasso Sea, 12, 194
Sargassum (*see* Brown algae)
Saturation, 311
Savonius rotor current meter, 282,
  285
Scandinavian Peninsula isostatic
  rebound, 17
Scotia Sea, 11, 12
Scuba, 274
Scyphozoa (*see also* Coelenter-
  ata):
  figure, 230
  synopsis, 230
Sea anemone (*see* Coelenterata)
Sea birds, 210
Sea cucumber (*see also* Echinoder-
  mata), 208
Sea-floor spreading, 43, 46–52
  rates of, 46–49
  and sedimentation, 268, 269
Sea gooseberry (*see* Ctenophore)
Sea lettuce (*see* Green algae)
Sea level, changes of, 52, 176, 177
Seamount, 30, 32, 46, 323
Seamount province, 30
  sedimentation associated with,
    262
Sea of Japan, 12
Sea of Okhotsk, 12
Sea snake (*see* Reptilia)
Sea squirt (*see* Tunicata)
Sea turtle (*see* Reptilia), 209

Sea urchin (*see also* Echinodermata), 208
Sea walnut (*see* Ctenophore)
Sea wasp (*see* Coelenterata)
Seawater:
  analysis, 72, 283
  buffering of, 63
  and chlorinity, 65
  color of, 85
  content, 57
  density, 74
  freezing, 74
  gasses dissolved in, 66
  hydration, 56
  major ions, 57
  mixing time, 2, 65, 116
  nutrients, 59
  origin, 60
  sources of, 60, 63
  trace elements, 57
  volume of, 4
Sediment (*see also* Marine sediment), 323
Sedimentary rocks, 313
Sedimentation:
  climatic control on, 265, 266
  continental margin, 52, 53
  continental rise, 260
  continental shelf, 268, 269
  magnetic polarity, 37, 39
  nearshore processes, 169–171,
    175
  rates for deep-sea sediments,
    268
Sediment sorting, 171, 172
Seiche, 131, 323
Seismic reflection measurements,
  279, 280
Seismic refraction measurements,
  280, 281
Seismology, 40–42
Semidiurnal tide, 134, 323
Sessile, 223, 323
Settling velocity of particles, 249
Set-up of sea surface by wind, 128
Severn River, 157, 159
Shallow-water wave, 120, 126, 127
Sharks, 210
Shelf break, 19, 52, 220, 323
Shelf edge (*see* Shelf break)
Shipek grab sampler, 275
Shrimps (*see also* Decapoda), 206
Silicate:
  clay buffer system, 63
  measurement of, 284, 287
  as a nutrient, 59, 60
Siliceous ooze, 257, 259, 260
Sill, 11, 188, 323
Silt, 249, 323
Sinking rate:
  phytoplankton, 227

Sinking rate (*cont.*)
sediment, 249
Size classification of sedimentary
materials, 249
Slicks, sea surface, 129
Solar-diurnal tide, 144
Solar radiation:
and primary productivity, 241
in the sea, 82–85
Solar spectrum, 82
Solstice, 136, 137, 146, 323
Solubility, 323
environmental changes on, 66
gasses, 66, 70
salts, 56
Sonar, 273
Sound speed of in seawater, 273
South China Sea, 12
Spectrophotometer, 283, 234
Spit, formation of, 173, 174
Sponge:
figure, 199
figure of spicule, 199
Spreading-rate calculations, 46–49
Spring tides, 134, 144, 145, 323
Squid (*see also* Cephalopoda),
204
Stability, 76, 78, 242, 298
Standing wave, 120, 131, 153, 323
Standing wave tides, 153, 154
Starfish (*see also* Echinodermata),
208
S.T.D., 281
Stenohaline, 232, 323
Stenothermal, 232, 323
Storm tide, 128, 129
Straits of Gibralter, 115
Stratigraphy of marine sediments,
264
Subantarctic Convergence, 110
Subantarctic Water, 98
Subarctic Water, 98
Sublittoral environment, 221–223,
323
organisms in the, 231
Submarine canyon, 21, 168, 175,
262, 263, 323
Submarine slumping, 262
Submarine volcanic component of
sediments, 248, 252, 253, 261
Submarine volcanoes, 46
Subtropical Convergence, 98
Subtropical High, 90, 323
Sulfate-reducing bacteria, 69
Sulfate reduction, 238
Summer solstice, 136, 137, 146
Sunlight in the sea, 215
Supersaturation, 67, 87
Supralittoral environment, 221,
324
Surf grass, 197

Surf zone, 324
Swallow float, 285
Swash, 163, 324
Swell, 124, 324
Syzygy, 144, 324

Tablemount, 30–32, 46, 324
Tasmania, 11
Taxonomy, 191
Tectonic barriers, 52
Temperature:
extremes in the world, 82
scales of, 303
seasonal variation with depth,
215
sea surface values, 84
and stability, 77
variations in the nearshore en-
vironment, 221
Terrigenous components of sedi-
ments, 248, 250, 257
Terrigenous sand, 257
Test (*see also* Protozoa), 198, 324
Thallophyta:
synopsis, 190
Thermal energy (*see also* Energy),
296, 303, 304
Thermal properties of water, 307
Thermocline, 78, 99
Thermometer:
crystal, 283
reversing, 282, 283
Tidal amplitude, 135, 147
Tidal bore, 157
Tidal constants, 148
Tidal constituents, 145–147, 150,
153, 324
Tidal currents, 153, 154
Tidal datum, 152
Tidal disruption hypothesis of the
earth's origin, 15
Tidal measurements, 158
Tidal power, generation of, 158,
159
Tidal prediction, 145, 150
machine, 150, 152
Tidal range, 134, 176, 221, 324
Tide, 324
description of, 133–135
equatorial, 142, 143
gauge, 158, 289
nearshore, 155, 157
producing force, 141
tropical, 142, 143
Titration technique, 71, 283
Tonga Trench, 30
Trace element, 57–59
analysis of, 71
Tractive force, 141, 324
Trade Winds, 104, 106, 190, 324

Transform faults, 28, 49
Trap for biological sampling, 291,
292
Trawl for biological sampling, 291
Trenches, 324
dimensions, 28–30
formation, 43–45, 49
heat flow in, 43
sedimentation in, 262
Trieste, 29, 274
Tristan da Cunha, 28
Trophic level, 235, 239, 240, 324
Tropical tides, 142, 143, 324
Tropic of Cancer, 12, 137
Tropic of Capricorn, 137
T-S diagram, 112, 113, 324
Tsientang River, 157
Tsunami, 127, 128, 324
Tsunami warning system, 128
Tunicata:
figure, 209
synopsis, 207, 210
Turbidite sediments, 260, 262,
263
Turbidity currents, 23, 248, 260–
263, 324
Turbulent mixing, 241–245

Ulva (*see* Green algae)
Undertow, 165
Underwater camera:
photographic, 272, 273, 287
television, 273, 274
Upwelling, 324
and biological productivity, 108,
238
and flushing of estuaries, 188
and surface winds, 102–104
Urochordata (*see* Tunicata)
U.S. National Oceanographic Data
Center, 270

Vaiont reservoir, 132
Varves, 324
in marine sediments, 263, 264
Vector diagram, 301
Velella (*see also* Coelenterata),
228, 230
Venus's girdle (*see* Ctenophore)
Vernal equinox, 137
Vertebrata, 210
Vitamins, 59
Viscosity, 324
and friction, 99
of seawater, 79
and sinking rate of marine or-
ganisms, 228
Volcanic glass in marine sedi-
ments, 253, 254

Volcanoes as a sediment source, 252, 253

Water masses, 110–116
Water type, 115
Wave, 324
  breaking, 163
  catastrophic, 127, 131
  celerity, 119
  classification of, 119, 120
  components, 119, 120
  deep-water, 119
  dispersion, 126, 127
  internal, 129
  in nearshore zone, 160–165
  node, 131
  progressive, 119, 120, 124, 153
  recorder, 289
  refraction, 161–163, 175
  shallow-water, 120
  standing, 120, 131
  swell, 124
  types of, 118
  wind, 121–123
Wave base, 51

Wave-cut platform, 32
Weathering products of rocks, 248
Weddell Sea, 116
Wegener, Alfred, 35–37
Weight (definition), 300
Westerlies, 91, 104
Westward intensification, 96, 108
West-Wind Drift, 98, 325
Whales:
  adaptations, 218, 234
  baleen, 205, 209, 210
  blue, 210
  evolution of, 214
  growth rate, 234
  killer, 209, 210
  sperm, 210
  whalebone, 210
Wind:
  effects of continental masses, 92, 93
  monsoon, 92, 93
  surface, 87, 90, 105
Wind-driven current, 99–109, 325
Wind set-up, 128, 325

Wind stress, 98–100, 106, 108
Wind waves, 121, 325
  characteristics of, 123
  formation of, 122
  turbulent mixing from, 241
Winter solstice, 137, 146
Worms:
  arrow, 207
  flat, 202, 203
  polychaete, 203, 205
  segmented, 205
Wyville Thompson Ridge, 225, 232

Yellow substances, 59

Zeolite minerals in marine sediments, 254, 264
Zonation of intertidal organisms, 224
Zooplankton, 220, 325
  copepods, 205, 228
  floating adaptations, 228
Zostera, 182, 196, 197

RSITY

ARY